THE GARRISON BURYING GROUND

THE GARRISON BURYING GROUND

A HISTORY OF CHEEKTOWAGA'S WAR OF 1812 CEMETERY

J A BUSCAGLIA

ANTIQUARIAN PRESS

Printed in the United States of America

ISBN: 978-1-7370531-4-9

Library of Congress Control Number: 2025916966

Antiquarian Press

P.O. Box 693

East Amherst, New York 14051

www.antiquarianpress.com

In Memory of Bob DeRoss

1965 - 2024

SPECIAL THANKS TO:

Joan Gadomski Klaich

"Sacred to the memory of those men who died of wounds received in the memorable fields of Chippewa, Bridgewater and Erie, – They rest in honor and deserve the gratitude of their country"

— LT COL RANNEY 4TH U.S. INFANTRY REGIMENT

CONTENTS

INTRODUCTION

This book has been written with deliberate simplicity—without affectation or literary embellishment—because its subject calls for truth, not flourish.

As an independent historian, I have spent years pursuing the faint trails of deeds, landmarks, and quiet remnants in archival corners often overlooked. The War of 1812 left its mark on this region not only through redrawn boundaries but also through lives shaped by settlement, conflict, and enduring resilience. My role has been less that of authoring a narrative than of unearthing one long buried in ledgers, maps, and the fading memory of place.

Readers will find no dramatics, no sweeping declarations within these pages. What is presented here is a faithful record—organized and interpreted with care, though never embellished beyond what the sources allow. If the style feels provincial, it is because the story itself belongs to this soil: to these towns, and to the families who have called them home.

This book is offered as a record, a reflection, and a gesture of preservation. May it inspire others to look more closely at the places they believe they know.

No apologies are made for its imperfections. They are the honest shadows of a work driven by curiosity, not polish. The purpose here is not refinement but revelation. And if the prose remains plain, it does so in service to that aim.

Let the record speak. Let the reader decide.

PREFACE

Growing up near the War of 1812 Cemetery, I have always been intrigued by its history and the stories surrounding it. I vividly recall my first encounter with the site as a child, captivated by the old cannon and the rows of white crosses. Over the years, I have heard numerous accounts about the General Hospital, which was purportedly located next to the burial ground, where all the soldiers interred there succumbed to wounds and disease. There are also stories of Native Americans, who fought on the American side, being buried there, as well as British prisoners of war.

I have decided to compile my thoughts and observations, incorporating all the knowledge I have gathered on this subject from various sources. My aim is to determine the true history of the site based on this information. This document reflects my personal opinion derived from my historical research and observations. It is not an academic paper but rather a first-person account of my findings.

Additionally, my research delves into the true extent of the burial ground. Before presenting my findings, I would like to extend my heartfelt gratitude to Cheektowaga Town Historian Maureen Gleason and David Przepiora of the Cheektowaga Historical Association. Without their invaluable assistance, this project would have been far more challenging.

Here are my findings.

1

IN THE BEGINNING

In December 1812, General Smyth ordered Colonel P. P. Schuyler to identify a suitable location for winter cantonment. At the time, Colonel George Izard commanded the Second Artillery, and due to the deplorable conditions at Flint Hill, it is likely he also sought an alternative site.

While Colonel Schuyler scouted for quarters to house the 5th, 12th, 13th, and 20th Consolidated Regiments, Colonel Izard may have been conducting a parallel search. Evidence suggests that Schuyler selected the northwest end of the Village of Williamsville for his cantonment, while Izard may have settled on the southeastern section along Eleven Mile Creek, near present-day Lehn Springs.

Both locations offered abundant fresh water supplied by natural springs.

A letter published in "Documentary History of the Campaign upon the Niagara Frontier In the year 1813 Part I; Cruikshank Pg. 19 dated December 12, 1812

Lieutenant Patrick McDonough—a subordinate in Colonel Izard's Second Artillery—adds valuable detail. McDonough wrote to his parents:

"We are now encamped in the woods, building huts which we expect to get into by the middle of next month. It is rather late in the season to be in tents. We have a very handsome situation on Eleven Mile Creek. The place is called after its owner Colonel Williams of New York. I hear he contemplated building his house next spring on the very ground on which we are building, and desired that not a piece of timber should be cut, as he wished it entirely shaded, but I can promise him that by that time there will not be a sapling standing within a mile of it. We marched from Black rock to this place on the 11th inst."

Based on McDonough's account, I believe the "very handsome situation on Eleven Mile Creek" refers to the Lehn Spring area in modern-day Williamsville, located approximately a half-mile from the cemetery as the crow flies.

After the artillery units left winter quarters and resumed operations, the huts were abandoned. Shortly thereafter, orders were issued to establish a hospital in preparation for the upcoming campaigns, and the abandoned huts were repurposed as hospital cantonments.

Documentary History of the Campaign upon the Niagara Frontier part IV; Cruikshank Pg. 341 from Poulson's American Daily Advertiser of Philadelphia – January 1813

Captain McKeon of the US Artillery arrived in New York on Tuesday last from the lines. He informs us that he left the Army in comfortable winter quarters at Eleven Mile River. (Eleven Mile Creek, Williamsville)

From "The Early Firm of Juba Storrs & Co."; Rev. Albert Bigelow

The declaration of war at once brought business to the young firm especially at Williamsville. They made a contract with the government for all the mill products they could furnish, for the army, and had work enough to do. There were no other mills nearer than Niagara Falls, except, as Lucius Storrs with somewhat poetic license was wont to say, one that wasn't bigger than a coffee mill. Then, in the winter of 1812-1813, barracks and a hospital were established about a mile from Williamsville up the creek, by cutting down trees and building huts in the woods.

Before these were built, one regiment of troops were cantoned in front of Williams house; afterwards another behind the house, down the creek. On the bank of the creek was a beautiful grove, and the engineers asked permission to cut down trees, to build barracks with. But this Mr. Caryl refused as unnecessary destruction. They replied, however, that his refusal made no difference – their asking was a mere matter of courtesy – and so they went to work in spite of his refusal, and cut down the trees.

Before the barracks were built, a temporary hospital had been established. In the tavern, Gen. Brown and Gen. Ripley and a British officer were sick, and the men were taken care of in the houses of the little village. And the saw mill of Juba Storrs & Co. furnished the lumber of which the boats were built that carried the troops across the river.

Thus on account of the mills and hospital, Williamsville was an important point in the war, and large rewards were offered to the Indians by the British for the burning of them. But they were afraid of being intercepted and cut off, so never attempted to burn them.

Thus began the War of 1812 Cemetery now located on Aero Drive. The cemetery was a bit southeast of the barracks and hospital and was the started on land belonging to the Holland Land Co. as noted in the title search done for the Town of Cheektowaga in 1985. The property was finally sold to Benjamin Barton in January 1816.

The troops used these quarters until about April 1813 when they joined the battle force attacking York (Toronto) on April 27, 1813. The cabins were pretty much abandoned until October 1813 when it was decided to use the cabins as a winter hospital. The U.S. Army Medical department describes the attack at York and the casualties as follows;

US ARMY MEDICAL DEPARTMENT Office of Medical History

On the twenty-seventh of April the attack on Little York took place, and after four days occupation of the town, the army with the wounded and sick was moved to Fort Niagara, where a tent hospital was organized two miles from the river. The ground was wet and low, and many of the wounded died from camp diarrhea and typhus fever, and in June, after the capture of Fort George, a general hospital was established at Lewistown, eight miles up the river, by advice of Surgeon Mann, Medical Director. This hospital consisted of two barns, besides a large number of hospital tents, and was well supplied with everything necessary for the comfort of the sick and wounded; which by the first of August had increased to nearly seven hundred. Here the patients improved very rapidly, the position of the hospital being salubrious, the tents policed with great care, and the diet being generous. In the army at Fort George, however, a most lamentable degree of sickness prevailed.

Confusion exists concerning the nature of the medical facilities in the Fort Niagara-Fort George area. Dr. Mann described a tent hospital located two miles east of Fort Niagara where the wounded from the attack on Toronto were cared for beginning on 8 May 1813 and also noted that about two hundred wounded were moved from Fort Niagara to Lewiston in mid-June 1813.

There was also a general hospital at Fort George, which was taken from the British in late May 1813 and held for several months. In June, some of the patients at Fort George were also moved to the higher, healthier facility at Lewiston. By August, however, more than one-third of the men still there were sick and, with half the medical staff too ill to work, only three surgeons and four mates were available to care for six hundred to seven hundred patients. Hospital stores were running short by September when Mann decided to move the general hospital from Fort George to a site near Buffalo and to send all 100 invalids ready for discharge with a surgeon's mate from Fort George to Greenbush.

During the last weeks of October 1813, it was decided to remove the sick and wounded from Lewiston because of the approaching winter and the dangerous situation. Two hundred and fifty men were transported to Fort Schlossler (Niagara Falls) from where they were taken up river by boat to Black Rock, then to Williamsville by wagon. There, the barracks which had been erected by Smyth's army was improved and used as a hospital, six patients to each cabin. Dr. James Mann was appointed surgeon in charge, later to be succeeded by Dr. Whitridge. Having settled his patients in Williamsville's wholesome accommodations, Mann set out to join General Wilkinson at his winter camp at Malone, leaving surgeon's mate Joshua Whitridge, a physician whose "services cannot be too highly appreciated," in charge.

From "Medical sketches of the campaigns of 1812, 13, 14"; James Mann

At Williamsville, the barracks which the preceding winter had been occupied by General Smythe's Division, were put into a comfortable state of repair, and improved during this winter as hospitals. These quarters were very extensive, and were, by much labor, rendered commodious for the number who composed this detachments of sick and convalescents. The wards were made warm; this was more important, as the men were destitute of clothing. The wards were not crowded, consequently were less liable to become infectious. Six patients only were placed in a room, wherefore it required but little labor and attention to preserve them in a state of cleanliness. This part of the duty the soldiers were, at all times, obliged punctually to execute. Those regulations which had respect to cleanliness were always scrupulously enforced within the hospitals under my direction.

At Williamsville before three weeks had expired, fifty of the convalescents were reported fit for duty, and were ordered to Fort Niagara. The barbarous deaths of most of these men, with the rest of the ill fated garrison, we have to lament, while emotions of just indignation irresistibly obtrude themselves, when reflecting upon the manner. Surprised and taken without resistance, most of the garrisons were deliberately murdered by the bayonet, after surrender.

Having deposed of the sick in quarters for winter, in conformity to the orders received from General Wilkinson, and finding their number daily decreasing; my services at Williamsville were believed to be no longer absolutely necessary. It became my duty, as it was my inclination, to pursue the army of General Wilkinson by land to Sackets Harbor, where information might be obtained to what point from Grenadier Island it had directed its course.

Permission was received from General Harrison, who had but just arrived from Detroit, and on whom the command of the post devolved, to leave the Niagara Frontier, and join the division of General Wilkinson. The charge of the hospital at Williamsville was resigned to Doctor Whiteridge, who, for assiduous attention to duty, was exceeded by no physician of the army.

The last week of November, I departed the Niagara Frontier, to follow the division of General Wilkinson.

Deaths at Williamsville from "A War of 1812 Death Register" Jack Bilow

Deaths 1812 – Oct – 1, Nov – 4, Dec -8 (total – 11)

Deaths 1813 – Jan – 9, Feb – 3, Mar – 4, Apr – 2, May – 0, Jun – 1, July – 0, Aug – 0, Sep – 1, Oct – 0, Nov – 2, Dec – 4 (total – 25)

Deaths at Williamsville 1814 – Jan – 11, Feb – 4, Mar – 11, Apr – 5, May – 8, Jun – 1, July – 2, (total – 42)

During the first half of 1814 there were a total of forty-two deaths listed at Williamsville. But after the battles of Chippewa, Lundy's Lane and Fort Erie the number of wounded and sick became immense.

Deaths at Williamsville 1814 Aug – 40, Sep – 75, Oct – 69, Nov – 36, Dec – 44 (total – 264)

Deaths at Williamsville 1815 Jan – 49, Feb – 28, Mar – 7, Apr – 3, May – 2, Jun – 4, Jul – 1, Oct – 1 (total – 95)

Other deaths no dates 27, (total 464)

"While Washington Burned: The Battle of Fort Erie 1814", Whitehorne Pgs. 50, 55, 56

Before the invasion, General Scott had periodically visited all the various regimental hospitals, warning the physicians that the campaign would bring busy times for them.[21]

21. [21]. LC, Amasa Troubridge Papers, ibid

His predictions were all too accurate. These field hospitals were supported by an existing hospital system. By 1814, general hospitals already were established in Burlington, Vermont, Plattsburgh, Malone and Greenbush, New York. The threat to facilities at Fort Schlosser convinced General Brown in late July to erect a permanent general hospital at Williamsville, about 12 miles east of Buffalo. Williamsville was the headquarters of the 5th Brigade of the New York State Militia. It had been used by the Army ever since General Smyth had established winter quarters there in 1812.

At that time, log barracks were built along the south side of the village's Main Street between Ellicott Creek and the Garrison Road. These barracks were converted into temporary medical facilities in October 1813. Patients from Lewiston were transported by boat to Fort Schlosser, then overland. Those capable of the trip were then sent to convalesce at the Greenbush hospital near Albany. The Buffalo area hospital was commanded by Dr. James Mann and Surgeons Mate Joshua B Whiteridge. [22]

22. [22]. Robert W Bingham "The History of Williamsville" Niagara Frontier Miscellany xxxv, (Buffalo Historical Society, 1947): pp. 100-101

The site for the new (General) hospital was selected on 29 July (1814) on the recommendation of senior surgeon, Doctor Ezekiah Bull. Thus, 90 acres and the stables of Raphael Cook's farm were leased for the construction of a general hospital. Ironically, the contract specified "no burying place in the premises" Despite this, a well kept series of mass graves along Aero Drive remain as the last vestiges of this hospital. These further provide evidence that having a burying ground adjacent to a medical facility was an accepted practice. The Williamsville Cemetery contains the remains of both US and British personnel. The British remains are in their own grave on one edge of the plot. The segregation of enemy and friendly dead has been a tacit, if not official, practice on both sides. The Williamsville facility was designated a general hospital in August, and Surgeon's Mate Joseph Lovell was assigned to its command. He was succeeded in 1815 by Doctor William Thomas.[23]

23. [23]. NA, R G 94 Entry 407, Box 90, Hogan Vouchers.

As soon as General Brown decided to concentrate patients at Williamsville, a vigorous building project was begun. At first, a large tent city was erected as indicated by the use of 3,000 board feet of timber for flooring 100 hospital tents and 12 loads of hay for bed ticking. Each tent could hold 16 to 18 men. In early September, General Brown, once more in command, embarked on his extensive effort to shelter the troops at Fort Erie.

Additional amounts of timber were sent across the Niagara to make tent flooring. Likewise, canvas was delivered for use by the besieged troops. Brown directed that the wounded be moved into permanent buildings so that every tent possible would be available for troops in the field. Captain John Larkin was named supervisory quartermaster for the construction of the Williamsville Hospital. He brought in skilled workmen from as far away as Rochester and Utica. Huge quantities of locally produced brick were purchased in Buffalo and hauled to Williamsville to help construct the hospital barracks. Subsequent purchases of glass and shingles to complete the buildings substantiate the permanent nature of the structures. Patients, among them Generals Scott and Riall, began being moved to the Williamsville site on 30 July.[24]

24. [24]. NA, R G 94 Entry 125, Box 134 Vouchers; Percy M Ashburn, "American Army Hospitals of the Revolution and the War of 1812" Bulletin of John Hopkins Hospital 46, pg 32

An additional general hospital was opened in Buffalo in July to accommodate the surge of wounded from the Battle of Chippewa. This was located at Sandy Town (The Front) 400 yards from Buffalo Creek. The wounded were brought to the site by boat then carried by litter the last few hundred yards. By 1 August, it held nearly 1,100 patients. The British raid on 3 August a few miles north at Scajaquada Creek demonstrated the hospital's vulnerability. Consequently, as many men as possible were removed to the growing Williamsville facility. Doctor Bull and Surgeons Mates Thomas and Lovell supervised the Williamsville facility. Surgeons Mate William E Horner remained at Buffalo caring for the small number of patients who could not be moved. Thereafter, the Buffalo Hospital served as the clearing center for casualties from Fort Erie sending patients to Williamsville as quickly as it could. Those who died there were buried in an adjacent graveyard, identified only years later when human remains were discovered.[25]

25. [25]. Percy M Ashburn, "History of the Medical department of the United States Army" (1929) pp. 33-34. Severance, ibid, p 247;

Several smaller regimental hospitals in the Buffalo and Black Rock areas had come in with their units. Despite their regimental affiliation, they supported all troops sent to the area. Buildings were leased to house these facilities. In addition, rooms in private houses were rented for the use of convalescent officers. Beginning in October, the sick and wounded at Williamsville who could travel were sent in a series of convoys to the hospital at Greenbush. Beginning on 8 November, everyone possible was moved to Williamsville. Most officers and about 80 critical patients remained in private houses around Buffalo. The unprecedented casualty rate experienced throughout the campaign combined with that of Plattsburgh and elsewhere to place heavy demands on the medical supply system.

Apothecary General Francis LeBaron at Albany advised his superiors in Philadelphia that the Williamsville Hospital "devoured" reserve stocks "like cormorants" so that by November nothing was left on hand.[26]

26. [26]. See annex K NA R G 94 Entry 125, Box 134; RG 94, Entry 225, Box 651, Medical Department, Francis LeBaron letter 14 November 1814 to Callender Irvine.

US ARMY MEDICAL DEPARTMENT Office of Medical History

In 1814, the hospital at Williamsville, New York, did not occupy the same quarters it had used the previous year, and patients were sheltered in tents while awaiting the completion of buildings designed specifically for use as a hospital. Progress on the construction was extremely slow, however, and General Izard wrote that "the jealousy and quarrels between surgeons and the Quartermaster's department of General Brown's division" were to blame. In early November 1814, the new building was still "far from ready for the reception of the sick and wounded."[16]

[16] Major General Izard to Secretary of War, 8 Nov 1814, Cruikshank, Documents, 4: 298.

Some of the construction may have been complete by 26 November, for General Izard referred to the Williamsville facilities at that time as an "extensive hospital establishment," but in early November, because of the inadequacies of the facility, General Izard, following a suggestion from the hospital's senior surgeon, ordered that as many as could tolerate the journey be sent on to Greenbush. Nevertheless, almost two thousand patients were in such poor condition that they had to remain behind in the tents at Williamsville. [17]

[17] Quote from Major General Izard to Secretary of War, 26 Nov 1814, Izard, Correspondence, p. 121; Major General Izard to Secretary of War, 8 Nov 1814, Cruikshank, Documents, 4: 298.

There was, for a time, also a hospital set up at Buffalo in 1814 in the tents left behind when camp there was broken. The Americans wounded in July's battle at Chippewa and some of the enemy wounded, probably those from the Battle of Lundy's Lane, were sent to Buffalo. Among the British patients were some so badly burned in the explosion of a powder magazine that their "faces and hands were so crisped that the skin peeled off like a baked pig." Among a number of American wounded who were rowed up the Niagara River in a flat-bottomed boat to Buffalo was General Scott himself. Only those in the poorest condition were retained at Buffalo for any length of time, however, and although some difficulties were experienced in obtaining transportation, all but eighty to ninety men too seriously injured to be moved were sent eastward, apparently within a few weeks, to the facility at Williamsville. The Buffalo facility was closed on 23 December 1814.[18]

[18] Quote from Eber D. Howe, "Recollections of a Pioneer Printer," Buffalo Historical Society Publications 9 (1906): 398; Clayton Tiffin to Major General Izard, 6 Dec 1814, R G 94, M 566, roll 59; Scott, Memoirs, 1: 147, 148; Report of Hospital Surgeon Lovell, 1 Aug 1814, Cruikshank, Documents, 4: 452-53; Louis L. Babcock, The Siege of Fort Erie, An Episode of the War of 1812 (Buffalo: Peter Paul Book Co., 1899), p. 44n; W. E. Horner, "Surgical Sketches," Medical Examiner and Record of Medical Science, new ser. 8 (1952): 761, 764, 768, 791; Jacob Brown to Secretary of War, 6 Jul 1814, in Herman Allen Fay, Collection of the Official Accounts, in Detail, of All the Battles Fought by Sea and Land Between the Navy and Army of the United States and the Navy and Army of Great Britain During the Years 1812, 13, 14, & 15 (New York, 1817), p. 210.

Because of the casual nature of some of the reports which survive from the time of the War of 1812, it is difficult to be sure of the nature of all of the hospitals serving the Army during that time. The Tilton report of 20 August 1814, for example, mentioned a return from a hospital surgeon, E. W. Bull, concerning the sick and wounded from Chippewa. It did not, however, make clear whether Bull was referring to men who became incapacitated during the battle of that name or to men hospitalized at some facility or facilities at or near Chippewa. If the former is true, however, the figures furnished Tilton by Bull (approximately 900 wounded, 300 sick) are, as Tilton commented, surprisingly high and probably included, as the Physician and Surgeon General commented, British casualties. It is difficult to believe that 300 men would fall ill as a result of one battle, and therefore it is likely that the reference was to the number of patients at a hospital. There is no record of a general hospital at Chippewa, however, but since the reporting physician was a hospital surgeon rather than a regimental surgeon, it is reasonable to assume that the Chippewa facility was a flying hospital.[19]

[19] James Tilton to Secretary of War, 20 Aug 1814, R G 107, M 221, roll 66; Major General Jacob Brown reported that 249 were wounded at the Battle of Chippewa ("Report of the Killed and Wounded of the Left Division Commanded by Major General Brown in the Action of 5th July 1814, on the Plains of Chippewa, Upper Canada, Brannan, Letters, p. 372.

Patients at Fort Erie, which was taken from the British in July, were cared for by regimental surgeons who received high praise in the summer of 1814. The men there suffered from the high rate of illness which was to be expected in the area, but by early August, they were in better health than had been expected. Not long thereafter, however, the British began an unsuccessful five-week siege of Fort Erie, during which a strong point blew up just as the British were about to take possession of it. The wounded left behind by the British, their faces, in some cases, "so fearfully disfigured, that the sight of them was sickening," were carried to the American hospital. The care two surgeons and three mates gave the men entrusted to them at Fort Erie [20] led the commanding general to comment upon their "active, humane, and judicious treatment of the wounded, both of the enemy and of our own."[21]

[20] Effner, "Adventures," pp. 51-52, quote from p. 52; E. P. Gaines to Secretary of War, 7 and 23 Aug 1814, Brannan, Letters, pp. 384, 399; Mann, Sketches, pp. 70, 111.

[21] E. W. Ripley to Brigadier General Gaines, 17 Aug 1814, Brannan, Letters, p. 392.

From Niagara Frontier #12 1965 pg's 51-57
"Jacob Porter Norton, A Yankee on the Niagara Frontier"

Jacob Norton was a young Lieutenant assigned to the 4th Infantry Regiment during the War of 1812. During his time of service he wrote letters home to his father and kept a diary of his activity.

Among his entries in his diary are;

November 14th 1814 "The 4th Reg't marched from our encampment near Buffalo, and took quarters in the Hospital Barracks, at eleven mile Creek in Williamsville, 11 miles from Buffalo."

June 1815, while writing his thoughts about fellow officers he had this to say about Lt Col Ranney; "The army burying ground at this place when we first came, was in a miserable situation,- Many of the graves were not more than half filled, and they had been dug promiscuously without regard to order, so that it was difficult in some places to distinguish the graves. Lt Col Ranney had them mounded up and clotted with sard, the stumps dug up, a fence built round it, a gate with an arch over the way, upon the arch was inscribed" – "Sacred to the memory of those men who died of wounds received in the memorable fields of Chippewa, Bridgewater and Erie, – They rest in honor and deserve the gratitude of their country"

On June 14th 1815 the 4th Infantry Reg't marched out of Williamsville beginning their journey home.

In an account as described in the *Buffalo Cemeteries: An account of the Burial-places of Buffalo, From the Earliest Times*, read before the Buffalo Historical Society,

February 4, 1879:

It would not be right to omit a notice of the soldier's burial place at Williamsville. — About six thousand of our army raised during the first year of the War of 1812 -15 and sent on to protect our frontier at Buffalo, went into winter quarters at Williamsville Village, situated eleven miles north of Buffalo. Their encampment at that place was just north of the main road and contiguous to the village on the extensive premises then owned by the enterprising merchant and milling firm of Juba Storrs & Co. This ground continued to be occupied by our soldiers more or less during the war. Sickness, as is usual in camp, prevailed among them; and some two or three hundred died and were buried on the grounds adjacent. Since then, the Village has spread, covering the ground where they were buried and long since not a vestige of a grave was left to be seen. There has been no one to look after or care for their bones when exhumed, as they often were, in excavating cellars or making improvements such as are necessary in a growing village.

There were quite extensive barracks built on the bank of the Eleven Mile Creek, a very healthy, eligible place, it being retired from the immediate scenes of conflict and about one mile above the Village. It continued to be our general hospital for sick and wounded soldiers during this war.

During the three years of the conflict, many of our soldiers died and were buried on the side of a nearby field. The ground that contains their remains is approximately half an acre in size. It lies on the southwesterly side of a public road; the Eleven Mile Creek running parallel and adjoining.

Two of the Village's townsmen, Col. John Bliss and John B. Evans, feeling an interest that this ground should never be disturbed or encroached upon, procured the title by a warranty deed from the owners, John Haskell and his wife, to themselves, dated August 6th, 1851, and duly recorded. They have both since deceased, leaving the title to their heirs. The number of our own soldiers buried on this ground is supposed to be three hundred or more; and in one retired corner lie nearly one hundred of our enemies who were wounded, taken prisoners and died. They captured many of the latter at the desperate sortie the British made on Fort Erie, August 14, 1814, and the explosion of its magazine. There were several hundred prisoners taken at this encounter; many of them were wounded most horribly, having been blown up when the magazine exploded.

On the second day after the sortie, I saw a number of wagon loads of those blackened and maimed British soldiers as they stopped in front of my father's house on their way to the hospital.

Recently, with one of the oldest residents of the town, Hon. T. A. Hopkins, I visited this burying place. There was not a slab or monument of any kind to be seen to designate the graves. Only the uneven hillocks marked the spot that contains the bones of our un-honored dead.

This ground is a little elevated from the surrounding land and road adjoining. On the border of two sides stand ten sugar maple trees, from one and one-fourth to two feet in diameter: all but one in a fine, healthy condition. I have no doubt they were planted there by the comrades of those whose bones now occupy this ground. The tenth tree is in a state of decay, and like many of the human race, is beginning to die at the top.

It is known only to a few living witnesses that this spot of ground contains the graves of some of our dead soldiers. Our government ought to erect at this beautiful, retired place a suitable monument to their memory and provide for removing thither the skeletons of others, when found in excavating in the village.

2

THE DEEDED HISTORY OF THE WAR OF 1812 CEMETERY

According to *Snake Hill: An Investigation of a Military Cemetery from the War of 1812,* the site selected for the new general hospital was chosen on July 29, 1814, based on the recommendation of Senior Surgeon Dr. Ezekiel W. Bull. A total of 90 acres, including the stables of Raphael Cook's farm, were leased for the hospital's construction. Interestingly, the lease contract explicitly prohibited the establishment of a burial site on the premises. The original wording reads: "No burying place be suffered on the premises."

It is likely this stipulation stemmed from the fact that Raphael Cook himself leased the land from the Frick family and was therefore unable to authorize permanent use of the land—such as a burial ground—beyond the term of his lease. A similar condition may have applied earlier when Mr. Frick hosted the first hospital on his own property during the summer of 1813, where Artillery Huts served as makeshift medical facilities.

The earliest documented acquisition of the land encompassing the military cemetery in Lot 7, Township 11, Range 7, occurred on October 4, 1808. This tract including Lots 6 and 7, totaling approximately 202 acres—was purchased by Frank Ludlow from the Holland Land Company.

Records indicate that Ludlow maintained ownership until January 11, 1816, when he forfeited the land due to nonpayment. There is no evidence suggesting Ludlow ever occupied or developed the land, which may explain why the U.S. Army selected it because it was vacant and unimproved.

On the same date, January 11, 1816, Wilhem Willink and his associates from the Holland Land Company sold the parcel—including the site of the cemetery—to Benjamin Barton. This transaction is officially recorded in the Clerk's Office (Liber 3, Page 133) and represents the first documented sale of the land containing the cemetery, although the deed makes no mention of a burial ground. See Figure 1 for the original deed.

Subsequent transactions include:

October 17, 1821: Barton sold the property to Thomas Coe (Liber 7, Page 207). See Figures 2 & 3.

July 4, 1851: William Coe transferred the land to John Haskell (Liber 172, Page 459). See Figure 4.

August 6, 1851: John Haskell sold Parcel A (the cemetery property) to John Bliss and John R. Evans (Liber 123, Page 19), with the stipulation that "the ground should not be disturbed or encroached upon." This marked the final private ownership of the cemetery. See Figure 5.

July 12, 1898: The estates of Bliss and Evans deeded Parcel A to the Buffalo Historical Society (Liber 833, Page 283), where it remained under the stewardship of the Buffalo History Museum until its transfer to the Town of Cheektowaga. See Figures 6, 7 & 8.

March, 1985 a quit-claim deed is signed conveying the cemetery (parcel A) from the Buffalo and Erie County Historical Society to the Town of Cheektowaga. (Liber 9444 page 450)

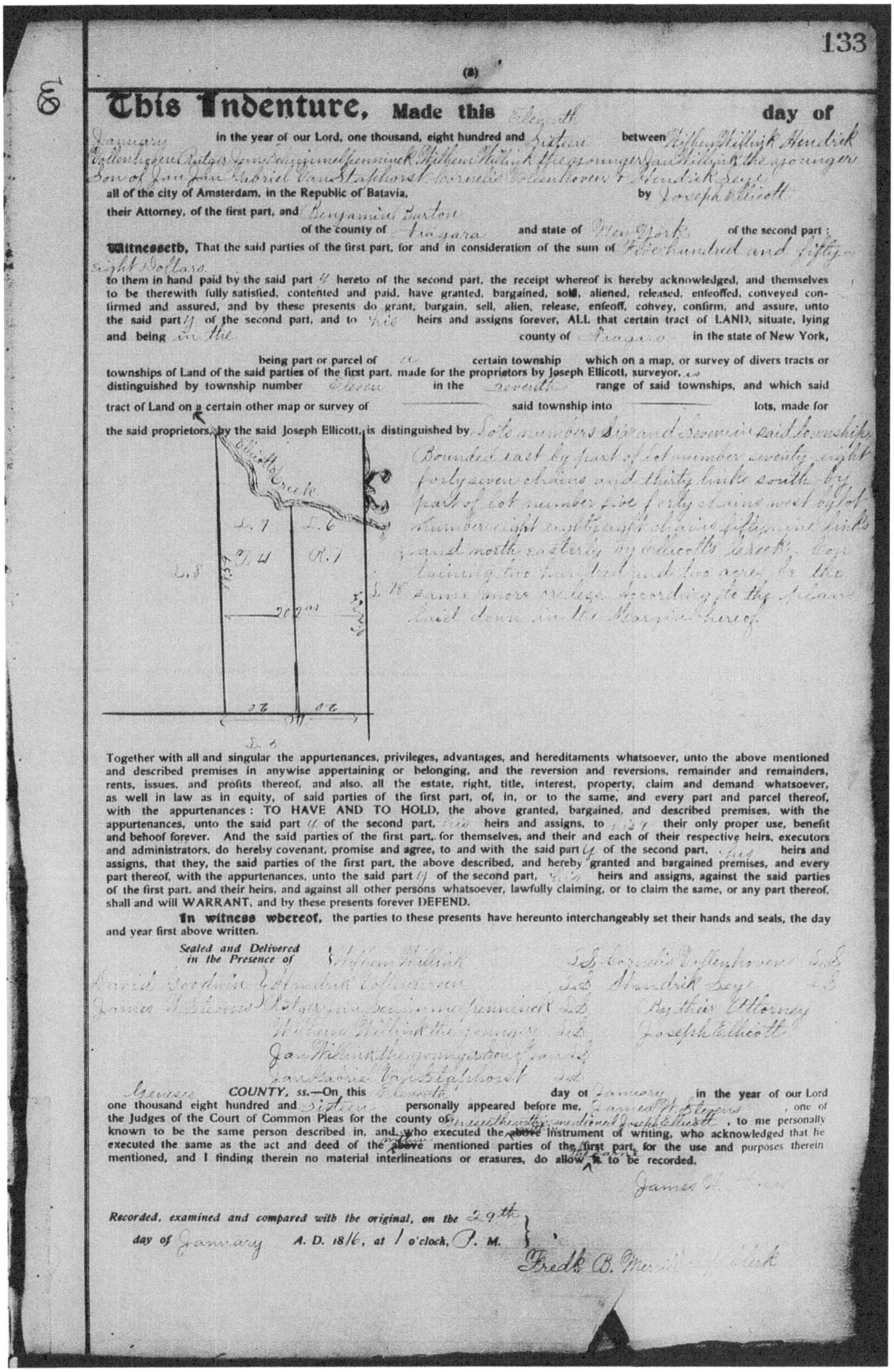

133

(a)

This Indenture, Made this Eleventh day of January in the year of our Lord, one thousand, eight hundred and Sixteen between Wilhem Willink Hendrik Vollenhoven Rutger Jan Schimmelpenninck Wilhem Willink the younger Jan Willink the younger Son of Jan Jan Gabriel Van Staphorst Cornelis Vollenhoven & Hendrik Seye all of the city of Amsterdam, in the Republic of Batavia, by Joseph Ellicott their Attorney, of the first part, and Benjamin Barton of the county of Niagara and state of New York of the second part:

Witnesseth, That the said parties of the first part, for and in consideration of the sum of Five hundred and fifty-eight Dollars to them in hand paid by the said part y hereto of the second part, the receipt whereof is hereby acknowledged, and themselves to be therewith fully satisfied, contented and paid, have granted, bargained, sold, aliened, released, enfeoffed, conveyed confirmed and assured, and by these presents do grant, bargain, sell, alien, release, enfeoff, convey, confirm, and assure, unto the said part y of the second part, and to his heirs and assigns forever, ALL that certain tract of LAND, situate, lying and being in the county of Niagara in the state of New York, being part or parcel of a certain township which on a map, or survey of divers tracts or townships of Land of the said parties of the first part, made for the proprietors by Joseph Ellicott, surveyor, is distinguished by township number Eleven in the Seventh range of said townships, and which said tract of Land on a certain other map or survey of said township into lots, made for the said proprietors, by the said Joseph Ellicott, is distinguished by Lots numbers Six and Seven in said township Bounded east by part of lot number seventy eight forty seven chains and thirty links south by part of lot number five forty chains west by lot number eight eighty eight chains fifty nine links and north easterly by Ellicotts Creek. Containing two hundred and two acres be the same more or less. According to the plan laid down in the Margin hereof.

Together with all and singular the appurtenances, privileges, advantages, and hereditaments whatsoever, unto the above mentioned and described premises in anywise appertaining or belonging, and the reversion and reversions, remainder and remainders, rents, issues, and profits thereof, and also, all the estate, right, title, interest, property, claim and demand whatsoever, as well in law as in equity, of said parties of the first part, of, in, or to the same, and every part and parcel thereof, with the appurtenances: TO HAVE AND TO HOLD, the above granted, bargained, and described premises, with the appurtenances, unto the said part y of the second part, his heirs and assigns, to his their only proper use, benefit and behoof forever. And the said parties of the first part, for themselves, and their and each of their respective heirs, executors and administrators, do hereby covenant, promise and agree, to and with the said part y of the second part, his heirs and assigns, that they, the said parties of the first part, the above described, and hereby granted and bargained premises, and every part thereof, with the appurtenances, unto the said part y of the second part, his heirs and assigns, against the said parties of the first part, and their heirs, and against all other persons whatsoever, lawfully claiming, or to claim the same, or any part thereof, shall and will WARRANT, and by these presents forever DEFEND.

In witness whereof, the parties to these presents have hereunto interchangeably set their hands and seals, the day and year first above written.

Sealed and Delivered in the Presence of

David Goodwin
James H. Stevens

Wilhem Willink L.S.
Hendrik Vollenhoven L.S.
Rutger Jan Schimmelpenninck L.S.
Wilhem Willink the younger L.S.
Jan Willink the younger son of Jan L.S.
Jan Gabriel Van Staphorst L.S.
Cornelis Vollenhoven L.S.
Hendrik Seye L.S.
By their Attorney
Joseph Ellicott

Genesee COUNTY, ss.—On this Eleventh day of January in the year of our Lord one thousand eight hundred and Sixteen personally appeared before me, James H. Stevens, one of the Judges of the Court of Common Pleas for the county of Genesee the within mentioned Joseph Ellicott, to me personally known to be the same person described in, and who executed the within instrument of writing, who acknowledged that he executed the same as the act and deed of the within mentioned parties of the first part, for the use and purposes therein mentioned, and I finding therein no material interlineations or erasures, do allow it to be recorded.

James H. Stevens

Recorded, examined and compared with the original, on the 29th *day of* January *A. D.* 1816, *at* 1 *o'clock,* P. *M.*

Fredk B. Merrill Clerk

Figure 1: Benjamin Barton Deed (Liber 3 p. 133) January 11, 1816

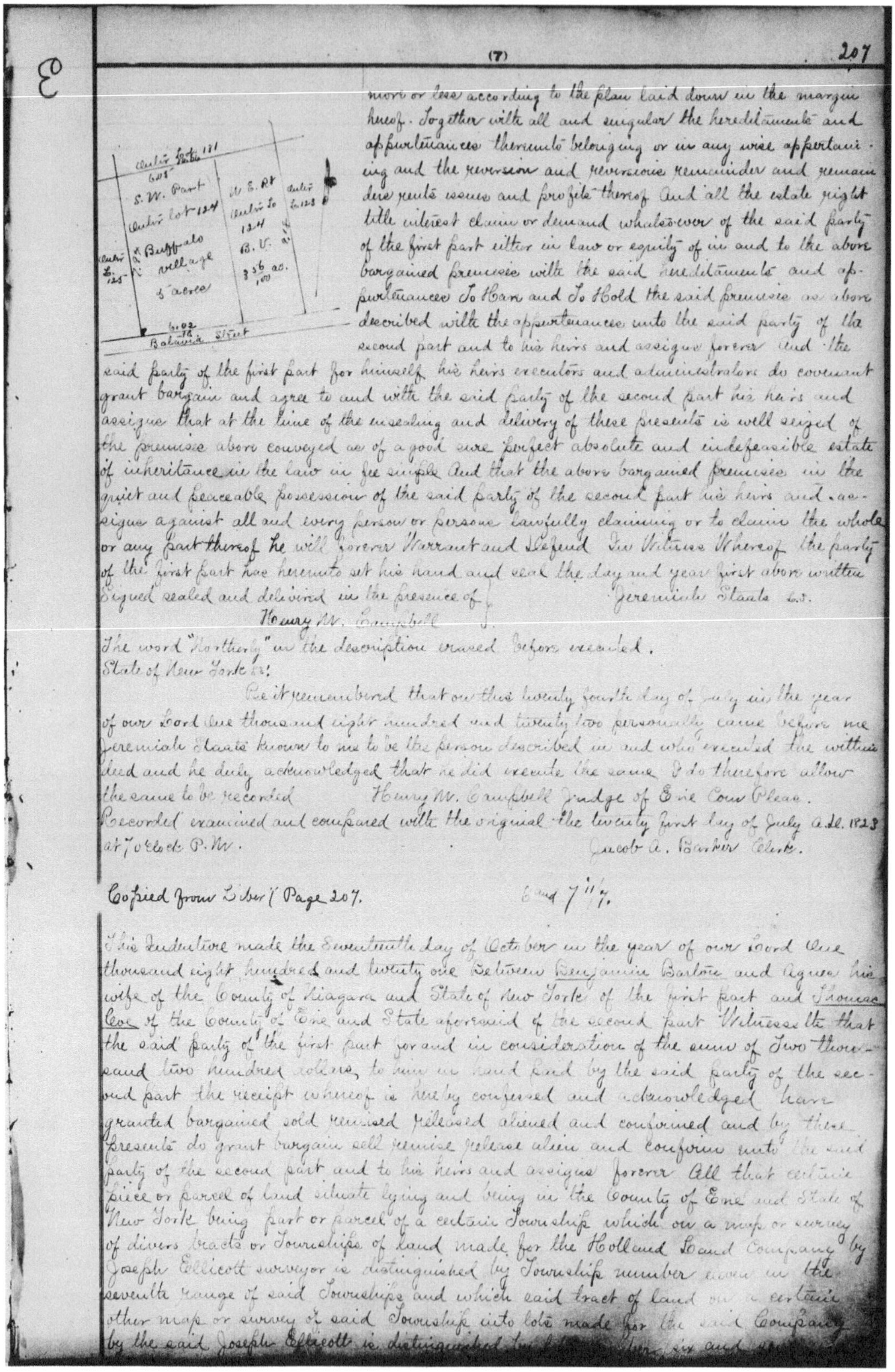

(7) 207

E

more or less according to the plan laid down in the margin hereof. Together with all and singular the hereditaments and appurtenances thereunto belonging or in any wise appertaining and the reversion and reversions remainder and remainders rents issues and profits thereof And all the estate right title interest claim or demand whatsoever of the said party of the first part either in law or equity of in and to the above bargained premises with the said hereditaments and appurtenances To Have and To Hold the said premises as above described with the appurtenances unto the said party of the second part and to his heirs and assigns forever And the said party of the first part for himself his heirs executors and administrators do covenant grant bargain and agree to and with the said party of the second part his heirs and assigns that at the time of the ensealing and delivery of these presents is well seized of the premises above conveyed as of a good sure perfect absolute and indefeasible estate of inheritance in the law in fee simple And that the above bargained premises in the quiet and peaceable possession of the said party of the second part his heirs and assigns against all and every person or persons lawfully claiming or to claim the whole or any part thereof he will forever Warrant and Defend In Witness Whereof the party of the first part has hereunto set his hand and seal the day and year first above written
Signed sealed and delivered in the presence of Jeremiah Staats L.S.
Henry M. Campbell

The word "Northerly" in the description erased before executed.

State of New York ss:

Be it remembered that on this twenty fourth day of July in the year of our Lord one thousand eight hundred and twenty two personally came before me Jeremiah Staats known to me to be the person described in and who executed the within deed and he duly acknowledged that he did execute the same I do therefore allow the same to be recorded Henry M. Campbell Judge of Erie Com Pleas.

Recorded examined and compared with the original the twenty first day of July A.D. 1823 at 7 o'clock P.M. Jacob A. Barker Clerk.

Copied from Liber 7 Page 207.

This Indenture made the Seventeenth day of October in the year of our Lord One thousand eight hundred and twenty one Between Benjamin Barton and Agnes his wife of the County of Niagara and State of New York of the first part and Thomas Coe of the County of Erie and State aforesaid of the second part Witnesseth that the said party of the first part for and in consideration of the sum of Two thousand two hundred dollars to him in hand paid by the said party of the second part the receipt whereof is hereby confessed and acknowledged have granted bargained sold remised released aliened and confirmed and by these presents do grant bargain sell remise release alien and confirm unto the said party of the second part and to his heirs and assigns forever All that certain piece or parcel of land situate lying and being in the County of Erie and State of New York being part or parcel of a certain Township which on a map or survey of divers tracts or Townships of land made for the Holland Land Company by Joseph Ellicott surveyor is distinguished by Township number eleven in the seventh range of said Townships and which said tract of land on a certain other map or survey of said Township into lots made for the said Company by the said Joseph Ellicott is distinguished ...

Figure 2: October 17th, 1821 Barton Sells Property Containing the Cemetery to Thomas Coe. (Liber 7 p .207) Page 1

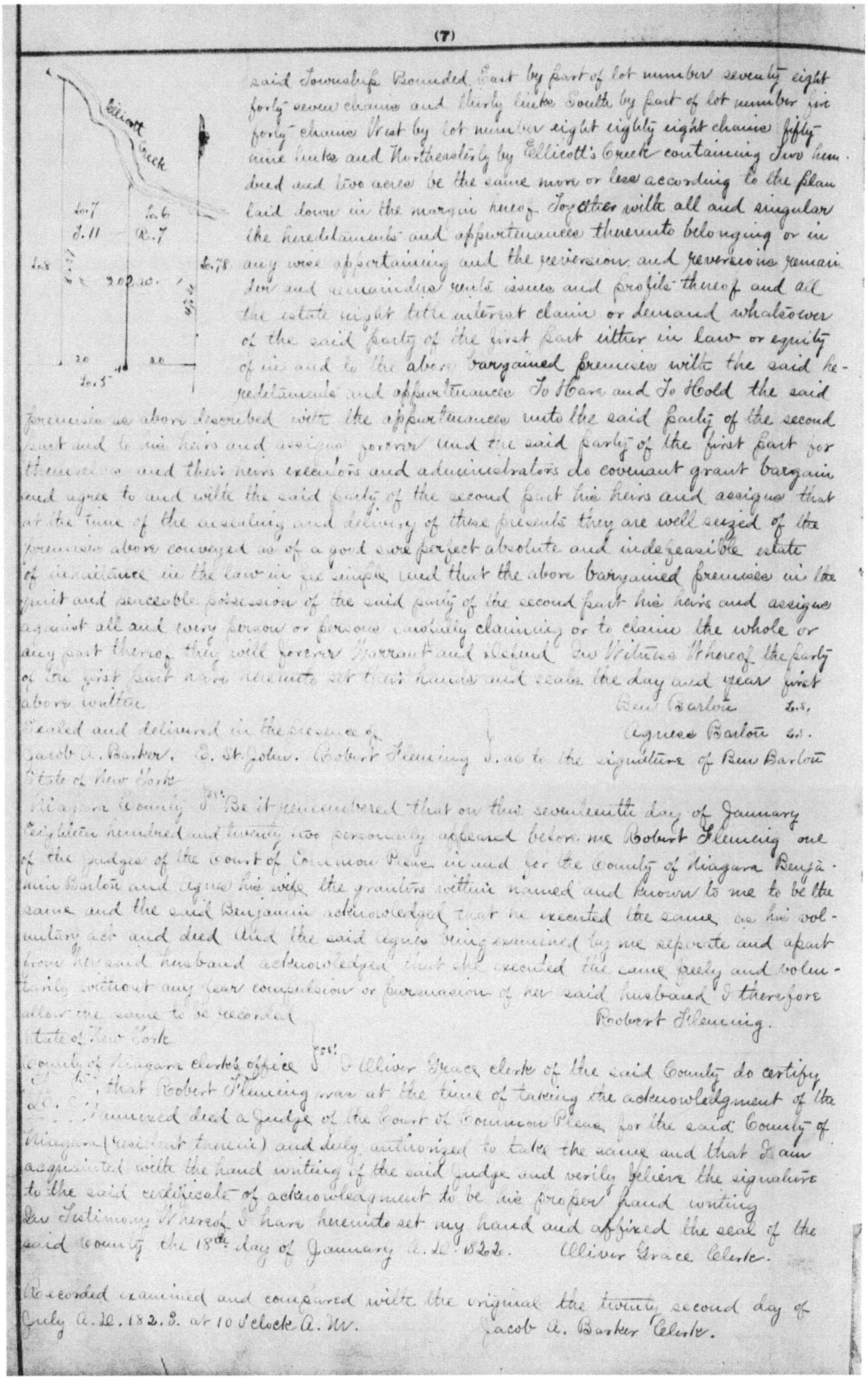

(7)

said Township Bounded East by part of lot number seventy eight forty seven chains and thirty links South by part of lot number five forty chains West by lot number eight eighty eight chains fifty nine links and Northeasterly by Ellicott's Creek containing two hundred and two acres be the same more or less according to the plan laid down in the margin hereof. Together with all and singular the hereditaments and appurtenances thereunto belonging or in any wise appertaining and the reversion and reversions remainder and remainders rents issues and profits thereof and all the estate right title interest claim or demand whatsoever of the said party of the first part either in law or equity of in and to the above bargained premises with the said hereditaments and appurtenances To Have and To Hold the said premises as above described with the appurtenances unto the said party of the second part and to his heirs and assigns forever And the said party of the first part for themselves and their heirs executors and administrators do covenant grant bargain and agree to and with the said party of the second part his heirs and assigns that at the time of the ensealing and delivery of these presents they are well seized of the premises above conveyed as of a good sure perfect absolute and indefeasible estate of inheritance in the law in fee simple and that the above bargained premises in the quiet and peaceable possession of the said party of the second part his heirs and assigns against all and every person or persons lawfully claiming or to claim the whole or any part thereof they will forever Warrant and Defend In Witness Whereof the party of the first part have hereunto set their hands and seals the day and year first above written

Ben Barton L.S.

Agnes Barton L.S.

Sealed and delivered in the presence of
Jacob A. Barker. E. St. John. Robert Fleming J. as to the signature of Ben Barton

State of New York
Niagara County ss. Be it remembered that on this seventeenth day of January eighteen hundred and twenty two personally appeared before me Robert Fleming one of the Judges of the Court of Common Pleas in and for the County of Niagara Benjamin Barton and Agnes his wife the grantors within named and known to me to be the same and the said Benjamin acknowledged that he executed the same as his voluntary act and deed And the said Agnes being examined by me separate and apart from her said husband acknowledged that she executed the same freely and voluntarily without any fear compulsion or persuasion of her said husband I therefore allow the same to be recorded

Robert Fleming.

State of New York
County of Niagara Clerk's office ss. I Oliver Grace clerk of the said County do certify that Robert Fleming was at the time of taking the acknowledgment of the annexed deed a Judge of the Court of Common Pleas for the said County of Niagara (resident therein) and duly authorized to take the same and that I am acquainted with the hand writing of the said Judge and verily believe the signature to the said certificate of acknowledgment to be his proper hand writing
In Testimony Whereof I have hereunto set my hand and affixed the seal of the said County the 18th day of January A.D. 1822. Oliver Grace Clerk.

Recorded examined and compared with the original the twenty second day of July A.D. 1822 at 10 o'clock A.M. Jacob A. Barker Clerk.

Figure 3: October 17th, 1821 Barton Sells Property Containing the Cemetery to Thomas Coe. (Liber 7 p .207) Page 2

WARRANTY DEED—COVENANT OF SEIZING (172) 459

Adams & White Company, Binders

This Indenture, Made the fourth day of July in the year of our Lord one thousand eight hundred and fifty one Between William Coe and Phebe Coe his wife of the Town of Cheektowaga in the County of Erie and State of New York. of the first part, and John Haskell of the Town of Amherst in the County of Erie and State of New York. of the second part.

Witnesseth, That the said party of the first part, for and in consideration of the sum of one thousand dollars. lawful money of the United States of America, to them in hand paid by the said party of the second part, the receipt whereof is hereby confessed and acknowledged, ha ve granted, bargained, sold, remised, released, aliened, conveyed and confirmed, and by these presents do grant, bargain, sell, remise, release, alien, convey and confirm unto the said party of the second part, and to his heirs and assigns forever.

All that certain piece or parcel of LAND, Situate in the Town of Cheektowaga, Erie County and State of New York being a part of Lot Number seven (7) in the eleventh (11th) Township and 7th range according to the Holland Land Company's Survey bounded and described as follows viz: Beginning at a point on the west line of said Lot No. 7 at the distance of about eighty chains and fifty links north of the southwest corner of said lot and in the center of the highway leading southeasterly through said lot; thence southerly along the west line of said Lot No. 7 thirty nine chains and fifty eight links to a post; thence easterly at a right angle with said west line and along the center of a rail fence eighteen chains and fifteen links to a post; thence northerly parallel to the said west line 4 chains and 30 links; thence westerly parallel to the south bounds 2 chains and 10 links to a post; thence northerly bounding on lands owned by the Heirs of Jacob Hershey and others 10 chains 99 links to the Eleven Mile Creek; thence down said Creek at the northwest corner of said Lot No. 7; thence southerly along the west line to the place of beginning, excepting from this conveyance a part of the same lot heretofore conveyed by John Coe to John Haskell hereby conveying about fifty acres be it more or less, according to the within plan, reference being thereto had.

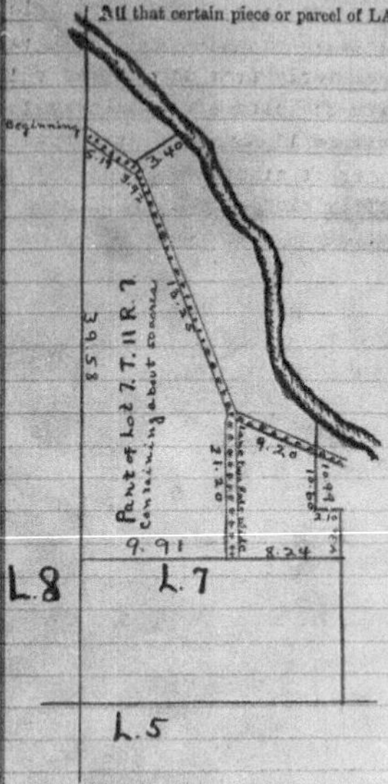

Together with all and singular the tenements, hereditaments and appurtenances thereunto belonging, or in anywise appertaining, and the reversion and reversions, remainder and remainders, rents, issues and profits thereof; and all the estate, right, title, interest, dower and right of dower, claim and demand, whatsoever of the said party of the first part, either in Law or Equity, of, in, and to the above bargained premises, with the said hereditaments and appurtenances.

To Have and to Hold, the said premises as above described, with the appurtenances, unto the said party of the second part, and to his heirs and assigns forever. And the said William Coe for himself and for his heirs, executors and administrators, does covenant, grant, bargain, and agree, to and with the said party of the second part, his heirs and assigns that at the time of the ensealing and delivery of these presents, he was well seized of the premises above conveyed, as of a good, sure, perfect, absolute and indefeasible estate of inheritance in the Law in Fee Simple. And that the above bargained premises, in the quiet and peaceable possession of the said party of the second part his heirs and assigns, against all and every person or persons lawfully claiming or to claim the whole or any part thereof he will forever Warrant and Defend.

In Witness Whereof, the party of the first part ha vehereunto set their hands and seals the day and year first above written.

Signed, Sealed and Delivered in Presence of

William Coe L.S.
Phebe Ann Coe L.S.

STATE OF NEW YORK, } ss.
Erie County. }

On this Fifteenth day of May 1855, before me the subscriber ~~personally~~ appeared William Coe and Phebe Ann his wife, and acknowledged that they had severally executed the within instrument. And the said Phebe Ann on a private examination, apart from her husband, acknowledged that she executed the within instrument freely, and without any fear or compulsion of her husband. And I further certify, that I know the persons who made the said acknowledgment to be the individuals described in, and who executed the within instrument.

John Hershey,
Justice of the Peace.

Recorded, Examined and Compared with the Original, the 25th day of September 1857, at 1 o'clock P. M. James H. Pratt Clerk.

Figure 4: July 4th, 1851, William Coe Sells Property Containing the Cemetery to John Haskell. (Liber 172 p.459)

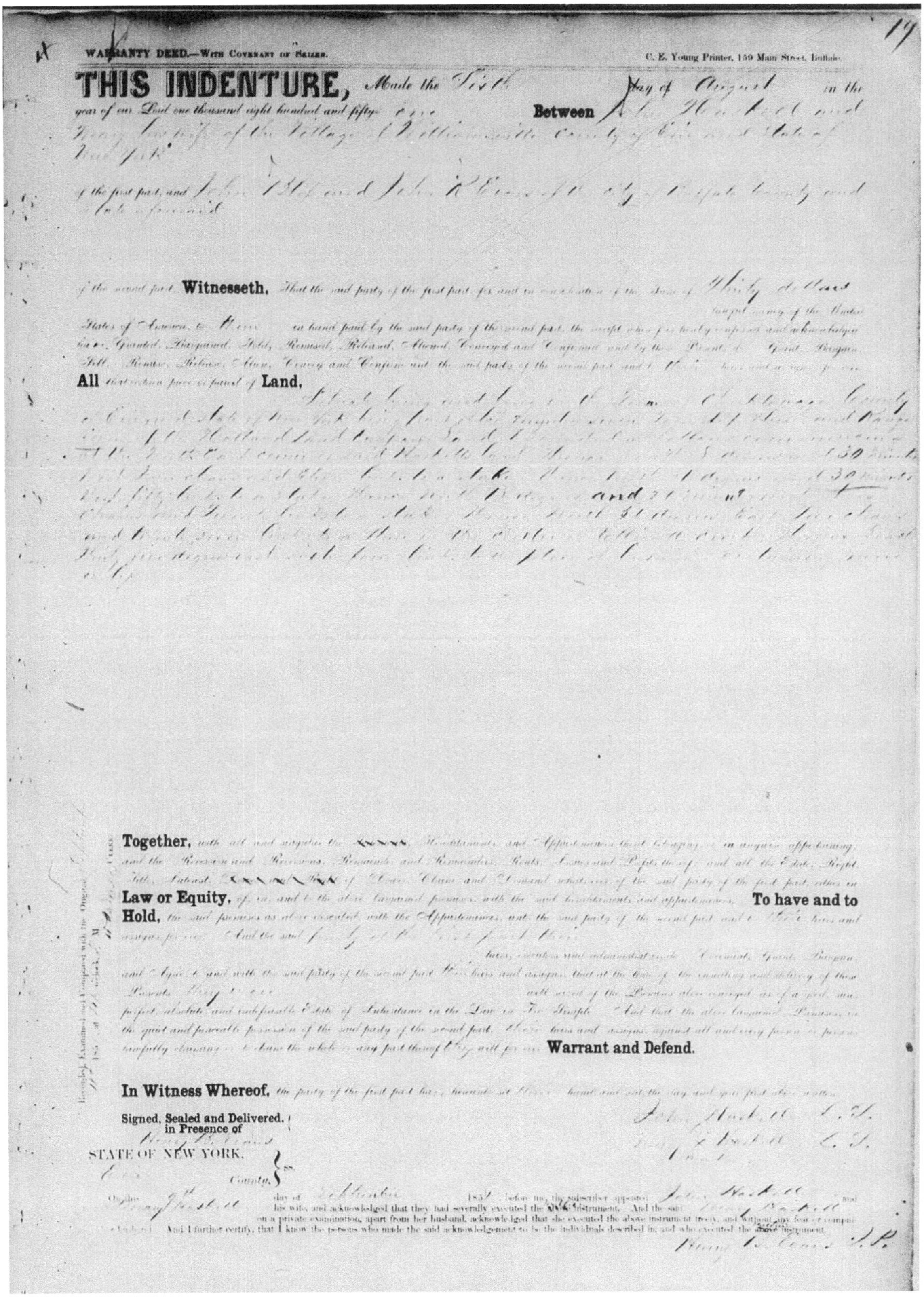
WARRANTY DEED.—WITH COVENANT OF SEIZEN. C. E. Young Printer, 159 Main Street, Buffalo.

THIS INDENTURE, Made the Sixth day of August in the year of our Lord one thousand eight hundred and fifty one Between John Haskell and ...

of the first part, and John Bliss and John R. Evans ...

of the second part Witnesseth, That the said party of the first part, for and in consideration of the Sum of Thirty dollars ... lawful money of the United States of America, to them in hand paid by the said party of the second part, the receipt whereof is hereby confessed and acknowledged, have Granted, Bargained, Sold, Remised, Released, Aliened, Conveyed and Confirmed ...

All that certain piece or parcel of Land, ...

Together, with all and singular the ... Hereditaments and Appurtenances thereunto belonging or in anywise appertaining ...

Law or Equity, of, in, and to the above bargained premises, with the said hereditaments and appurtenances. To have and to Hold, the said premises as above described, with the Appurtenances, unto the said party of the second part ... heirs and assigns forever. ...

... will forever Warrant and Defend.

In Witness Whereof, the party of the first part have hereunto set their hand and seal the day and year first above written.

Signed, Sealed and Delivered, in Presence of

John Haskell L.S.

STATE OF NEW YORK, Erie County, ss.

Henry R. Evans J.P.

Figure 5: August 6th, 1851, John Haskell Sells Cemetery Only to John Bliss and John R. Evans. (Liber 123 p. 19)

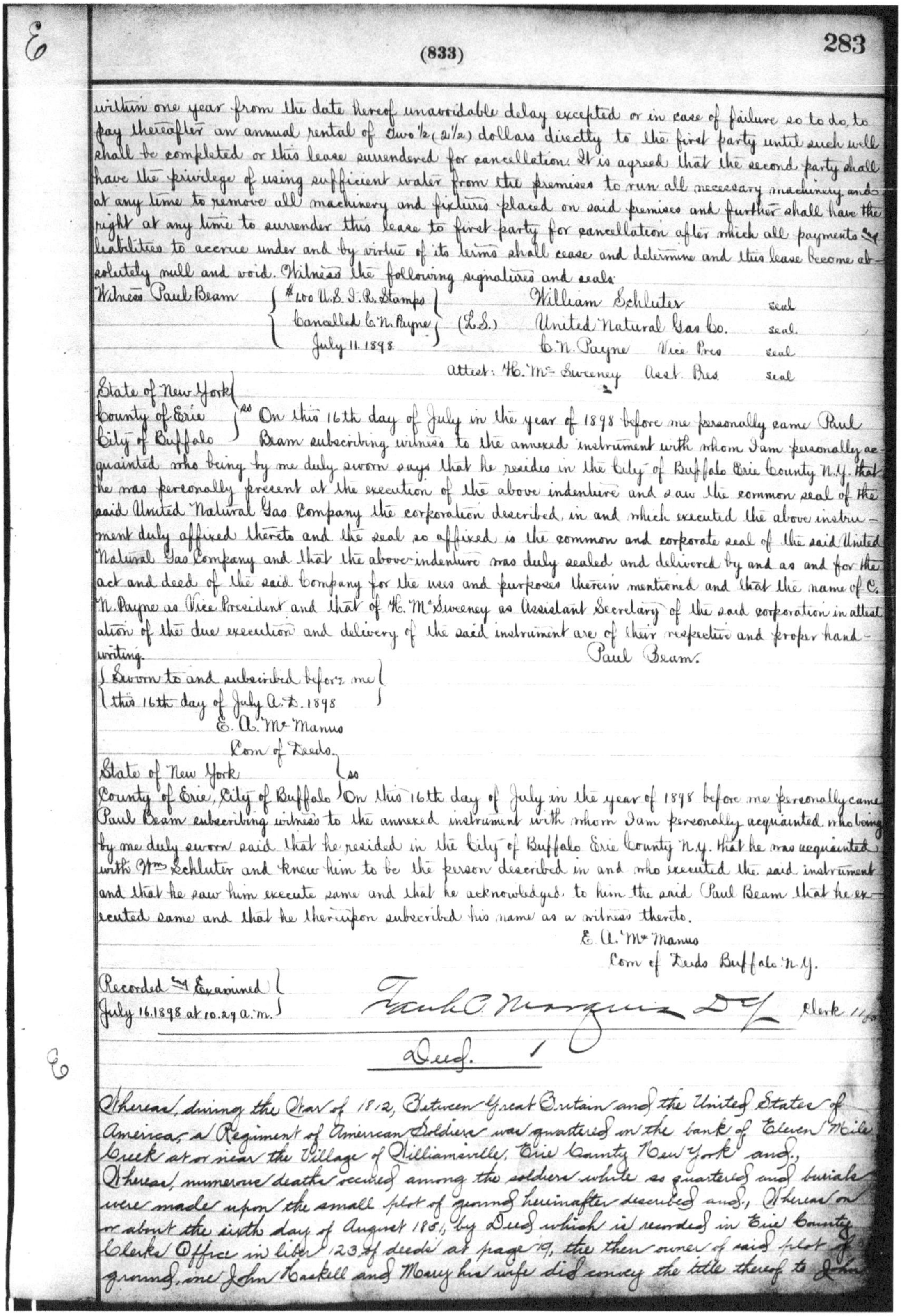
E

(833) 283

within one year from the date hereof unavoidable delay excepted or in case of failure so to do, to pay thereafter an annual rental of Two ½ (2½) dollars directly to the first party until such well shall be completed or this lease surrendered for cancellation. It is agreed that the second party shall have the privilege of using sufficient water from the premises to run all necessary machinery and to at any time to remove all machinery and fixtures placed on said premises and further shall have the right at any time to surrender this lease to first party for cancellation after which all payments & liabilities to accrue under and by virtue of its terms shall cease and determine and this lease become absolutely null and void. Witness the following signatures and seals.

Witness Paul Beam { $100 U.S. I. R. Stamps Cancelled C. N. Payne July 11. 1898. } (L.S.) William Schluter seal
United Natural Gas Co. seal
C. N. Payne Vice Pres seal
Attest: H. McSweeney Asst. Sec. seal

State of New York }
County of Erie } ss On this 16th day of July in the year of 1898 before me personally came Paul
City of Buffalo } Beam subscribing witness to the annexed instrument with whom I am personally acquainted who being by me duly sworn says that he resides in the City of Buffalo Erie County N.Y. that he was personally present at the execution of the above indenture and saw the common seal of the said United Natural Gas Company the corporation described in and which executed the above instrument duly affixed thereto and the seal so affixed is the common and corporate seal of the said United Natural Gas Company and that the above indenture was duly sealed and delivered by and as and for the act and deed of the said Company for the uses and purposes therein mentioned and that the name of C. N. Payne as Vice President and that of H. McSweeney as Assistant Secretary of the said corporation in attestation of the due execution and delivery of the said instrument are of their respective and proper handwriting.

Paul Beam.

{ Sworn to and subscribed before me
{ this 16th day of July A.D. 1898 }
E. A. McManus
Com of Deeds.

State of New York } ss
County of Erie, City of Buffalo } On this 16th day of July in the year of 1898 before me personally came Paul Beam subscribing witness to the annexed instrument with whom I am personally acquainted who being by me duly sworn said that he resided in the City of Buffalo Erie County N.Y. that he was acquainted with Wm Schluter and knew him to be the person described in and who executed the said instrument and that he saw him execute same and that he acknowledged to him the said Paul Beam that he executed same and that he thereupon subscribed his name as a witness thereto.

E. A. McManus
Com of Deeds Buffalo N.Y.

Recorded & Examined {
July 16. 1898 at 10.29 a.m. } Frank C. Marquis DY Clerk

Deed. 1

E

Whereas, during the War of 1812, between Great Britain and the United States of America, a Regiment of American Soldiers was quartered on the bank of Eleven Mile Creek at or near the Village of Williamsville, Erie County New York and, Whereas, numerous deaths occurred among the soldiers while so quartered and burials were made upon the small plot of ground hereinafter described and, Whereas on or about the sixth day of August 1851, by Deed which is recorded in Erie County Clerks Office in liber 123 of deeds at page 19, the then owner of said plot of ground, one John Haskell and Mary his wife did convey the title thereof to John

Figure 6: July 12th, 1898, the Estates of John Bliss and John R. Evans Deeded the Cemetery (Parcel A) to the Buffalo Historical Society. (Liber 833 p. 283) Page 1

284 (833)

E

Bliss and John R. Evans with the intent and purpose of dedicating said plot of ground perpetually to the memory of said deceased soldiers and, Whereas said John Bliss and John R. Evans have both since died, leaving the undersigned parties of the first part their sole heirs at law and, Whereas, The Buffalo Historical Society of Buffalo, Erie County, New York is willing to take the title to said burial ground and to give the same proper care and attention, and in the very nature of things appears to be the Ideal custodian for such a little burial ground associated as it is with the history of Erie County. Now therefor, with the sole intent and purpose of dedicating said premises for the sole purpose they are now used and to place the record title thereof in a suitable custodian where the same will be held perpetually sacred to the memory of said deceased soldiers of the War of 1812, We the undersigned Rachel L. Evans, Francis F. Evans, Mairam H. Evans, John T. Evans and Georgiana H. Evans his wife and Martha C. Evans, sole surviving children and heirs at law of John R. Evans, late of Erie County N.Y. now deceased and John Horace Bliss of Erie Pennsylvania sole surviving heir at law of John Bliss late of Erie County N.Y. now deceased for a good and valuable consideration to them in hand paid the receipt of which is hereby acknowledged have bargained, sold remised and quit claimed and by these presents do bargain, sell, remise, and quit claim unto the Buffalo Historical Society of Buffalo, Erie County New York and its successors forever and in consideration of one Dollar ($1.00).

All that certain tract or parcel of land situate lying and being in the Town of Cheektowaga, County of Erie and State of New York and being part of lot number seven (7) of the Holland Land Companys land and bounded as follows; Commencing at the northeast corner of said Haskells land, thence south eight (8) degrees and thirty (30) minutes west four (4) chains and eleven (11) links to a stake, thence north eighty one (81) degrees thirty (30) minutes west fifty (50) links to a stake, thence north eighteen (18) degrees and twenty (20) minutes west three (3) chains and twenty (20) links to a stake, thence north fifty (50) degrees east two (2) chains and twenty seven (27) links to a stone in the center of Ellicott Creek, thence south thirty five (35) degrees east sixty four (64) links to the place of beginning, containing more or less together with the appurtenances and all the estate and rights of the said parties of the first part, the said heirs of John Bliss and John R. Evans deceased in and to the premises above described. To Have and to Hold the above bargained premises unto the said Buffalo Historical Society and its successors forever, said premises to remain and be perpetually a cemetary and resting place for the dead soldiers of the War of 1812.

In Witness Whereof, the parties of the first part have hereunto set their hands and seals this 12th day of July, A.D. 1898.

Rachel L. Evans (L.S)
Miriam Hunt Evans (L.S)
John T. Evans (L.S)
Georgiana H. Evans (L.S)
Martha C. Evans (L.S)
Jno. H. Bliss (L.S)

State of New York
County of Erie } ss
City of Buffalo.

On this 12th day of July 1898 before me the subscriber personally came Rachel L. Evans, Miriam Hunt Evans, John T. Evans, Georgiana H. Evans, Martha C. Evans to me known to be the same persons named in and who executed the foregoing instrument and they severally

Figure 7: July 12th, 1898, the Estates of John Bliss and John R. Evans Deeded the Cemetery (Parcel A) to the Buffalo Historical Society. (Liber 833 p. 283) Page 2

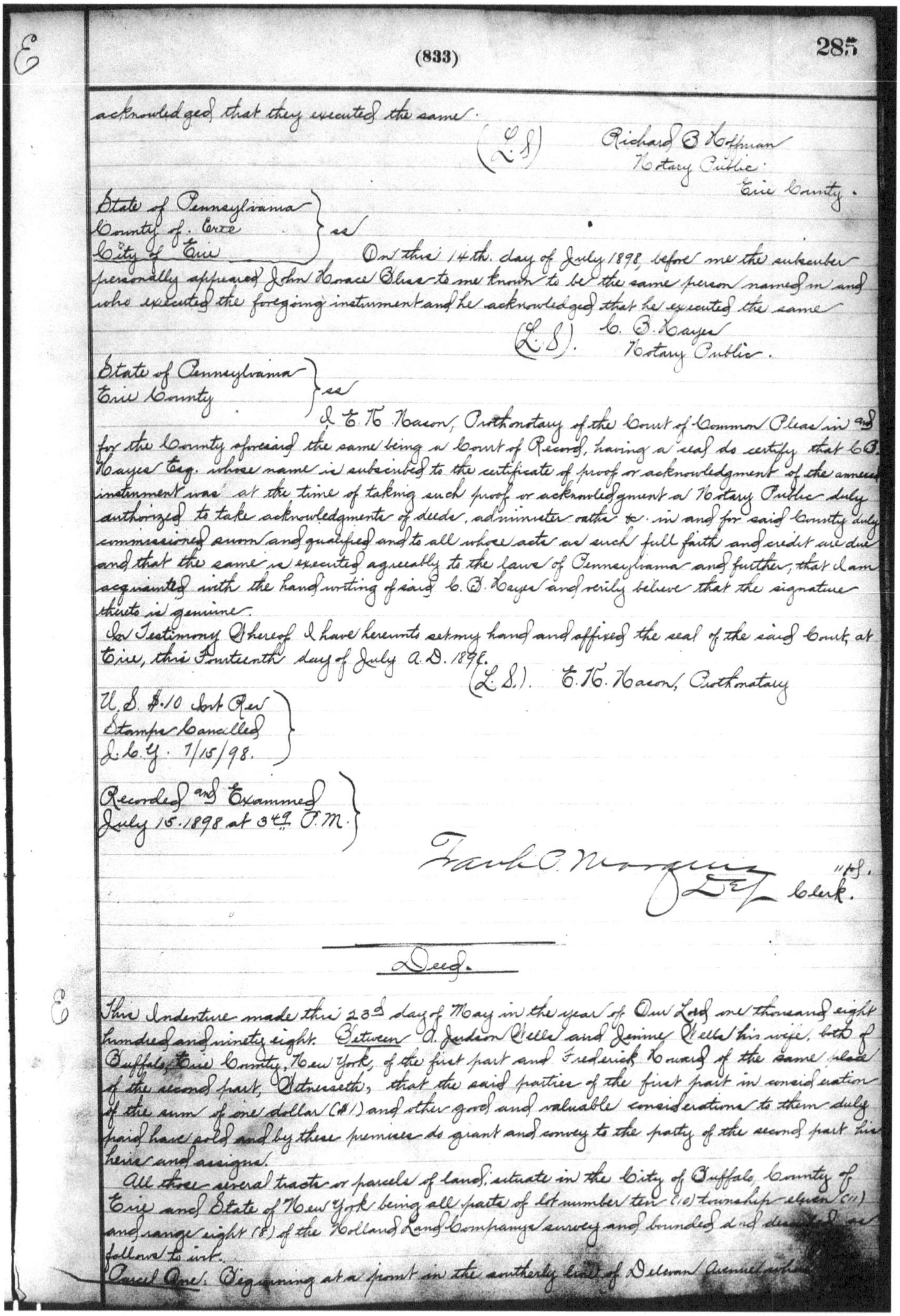

E (833) 285

acknowledged that they executed the same.

(L.S.) Richard B. Hoffman
Notary Public.
Erie County.

State of Pennsylvania }
County of Erie } ss
City of Erie }

On this 14th. day of July 1898, before me the subscriber personally appeared John Horace Bliss to me known to be the same person named in and who executed the foregoing instrument and he acknowledged that he executed the same.

(L.S.) C. B. Hayes
Notary Public.

State of Pennsylvania } ss
Erie County }

I, E. H. Nason, Prothonotary of the Court of Common Pleas in and for the County aforesaid the same being a Court of Record, having a seal do certify that C.B. Hayes Esq. whose name is subscribed to the certificate of proof or acknowledgment of the annexed instrument was at the time of taking such proof or acknowledgment a Notary Public duly authorized to take acknowledgments of deeds, administer oaths &c. in and for said County duly commissioned sworn and qualified and to all whose acts as such full faith and credit are due and that the same is executed agreeably to the laws of Pennsylvania and further, that I am acquainted with the handwriting of said C. B. Hayes and verily believe that the signature thereto is genuine.

In Testimony Whereof I have hereunto set my hand and affixed the seal of the said Court, at Erie, this Fourteenth day of July A.D. 1898.

(L.S.) E. H. Nason, Prothonotary

U.S. $.10 Int. Rev. Stamps Cancelled J.C.Y. 7/15/98.

Recorded and Examined July 15. 1898 at 3 49 P.M.

Frank C. Morgan, Dep. Clerk.

Deed.

E

This Indenture made this 23d day of May in the year of Our Lord one thousand eight hundred and ninety eight. Between A. Judson Wells and Jennie Wells his wife, both of Buffalo, Erie County, New York, of the first part and Frederick Howard of the same place of the second part, Witnesseth, that the said parties of the first part in consideration of the sum of one dollar ($1) and other good and valuable considerations to them duly paid have sold and by these presents do grant and convey to the party of the second part his heirs and assigns.

All those several tracts or parcels of land, situate in the City of Buffalo, County of Erie and State of New York being all parts of lot number ten (10) township eleven (11) and range eight (8) of the Holland Land Company's survey and bounded and described as follows to wit.

Parcel One: Beginning at a point in the southerly line of Delaware Avenue where

Figure 8: July 12th, 1898, the Estates of John Bliss and John R. Evans Deeded the Cemetery (Parcel A) to the Buffalo Historical Society. (Liber 833 p. 283) Page 3

3

CORRESPONDENCE AND CONSEQUENCE: THE SHELDON-HOPKINS LETTERS & PAPERS

In 1876, the United States War Department formally commissioned Alexander J. Sheldon, Librarian of the Grosvenor Library in Buffalo, New York, to prepare a report identifying the burial locations of soldiers who served in the War of 1812. The initiative sought to "collect the scattered ones and raise a memorial to them; and if there are any places where numbers are buried, they will be enclosed and taken care of as other National Cemeteries are." This directive reflected a growing national interest in honoring the sacrifices of early American servicemen and preserving the historical record of their contributions.

To advance this effort, Mr. Sheldon reached out to Timothy A. Hopkins, a respected local historian and son of Brigadier General T. S. Hopkins, seeking his assistance in locating these burial sites. Tragically, Mr. Sheldon passed away shortly after composing the letter, and the project remained dormant for more than two decades. The loss of Sheldon, a key figure in Buffalo's intellectual and civic life, left a noticeable void in the initiative's momentum.

The effort was revived on July 12, 1898, when the estates of John Bliss and John R. Evans deeded Parcel A the cemetery in question to the Buffalo Historical Society. This transfer marked a significant step toward preserving the legacy of those interred and signaled renewed public commitment to commemorating the region's military heritage. The cemetery, long overlooked and vulnerable to neglect, was now poised to receive the recognition and care befitting its historical importance.

Further context surrounding the War Department's request is provided by an article from the *Buffalo Courier*, dated Monday, January 24, 1876, which was independently located by the author. The original correspondence between Mr. Sheldon and Mr. Hopkins is preserved in the personal collection of Nelson S. Hopkins, housed at the Buffalo History Museum. These letters offer a rare glimpse into the early efforts to document and protect the final resting places of War of 1812 veterans.

Additional insights are drawn from the handwritten notes of Mr. T. A. Hopkins, in which he reflects on his personal experiences and memories of his father, Brigadier General T. S. Hopkins, who served in Williamsville during the War of 1812. On page 9 of these notes, Mr. Hopkins references a second burial ground within the village and provides a general description of its location. He also documents the positions of the Infantry Cantonment, Artillery Cantonment, and Hospital Cantonment. A transcription of this page, prepared by the author, is included in this chapter.

Together, these sources form a compelling narrative of historical preservation, civic responsibility, and familial remembrance. They underscore the importance of local efforts in safeguarding national memory and highlight the enduring impact of individual contributions to historical scholarship. The rediscovery and documentation of these burial sites not only honor the fallen but also enrich our understanding of the region's role in shaping early American military history.

The Dead Soldiers of 1812.

Alexander J. Sheldon, Esq., has received a communication from the War Department, requesting information as to the location of graves of soldiers of the war of 1812, in this vicinity, from which it would appear that the government at last purposes to pay some tribute of distinction to the long neglected remains. Mr. Sheldon is to state what general plan he has in view in relation to the disposition. He desires to make as full a report as possible, and is confident that if the Park Commissioners, Common Council, and citizens will manifest a generous interest in the matter, the Department will meet them in the erection of suitable monuments at Williamsville, Flint Hill, and on "The Front."

It is well known that large numbers of the dead soldiers were buried throughout and about the city. Some have been removed to the cemeteries, and the bones of others continue to be exhumed as city improvements advance. The principal places of interment are known to Mr. Sheldon, but there are doubtless many graves which he would be unable to ascertain unless informed by those who chance to be aware of their location. There are a number in the new park, the place being marked by two willows planted by Dr. Daniel Chapin, of whose farm the ground was then a part. On the banks of Niagara River and Scajaquada Creek, in various remote places, and at Williamsville large numbers were buried while at "Sandy Town" trenches were filled with the bodies. Any of our citizens who may possess information which would be of value in the matter should consider it their duty to communicate with Mr. Sheldon.

Buffalo Courier dated Monday, Janurary 24, 1876

Copy of ms. In possession of Nelson S. Hopkins August 3, 1938
Transcription of letter from Alex J. Sheldon to Timothy A. Hopkins Esq. on January 15, 1876

93 Folsom Avenue
Buffalo Jary 15, 1876

Timothy A. Hopkins Esq

Dear Sir

The War Department has sent me a request that I give a report as soon as possible as regards the burial places of Soldiers of the War of 1812. Some arrangement will doubtless be made to collect the scattered ones and raise memorials to them, and if there {are} any places where numbers are buried, they will be enclosed & taken care of ad other National Cemeteries are.

I believe in your Village there is a spot of land which was devoted during the war for that purpose. Will you at your earliest convenience have the kindness to tell me about how many were buried in your Village; also how, whether in coffins or merely hurled into pits as in many places with their clothes on.

Also are there scattered graves of Officers or Soldiers & if so can they be identified, and about how many are they?

Was the land given that they are buried in or by what tenure are the dead allowed there?

Any information you can give me will be thankfully received and when I collect all that I can, I shall submit the letters with a report to the Department. I am of the opinion that if all of the dead Soliders in & about your place can be collected together, that their last resting place will be made a thing of beauty an ornament to it; though it may be we have to wait, till after we have a very different House of Representatives Ere we can obtain the money for it. Perhaps the residents would join in & aid the matter &at the same time there might be some sort of a dedication of grounds.

Yours very truly
Alex Sheldon

Copy of ms. In possession of Nelson S. Hopkins August 3, 1938
Transcription of letter to Alex J. Sheldon Esq. in reply to his letter of Jan 15, 1876

Williamsville Feb. 10, 1876

Alex J. Sheldon Esq

Dear Sir

In reply to yours as the Condition, Title be of the Grounds in and about our Village used for the burial of Soldiers during the War of 1812.

There were 6,000 troops ordered to Williamsville for the winter quarters in December following the Declaration of War, and more or less remained here until about the Close of Hostilities.

The main body were located on the north side of Main Street in the Village, their burial place was in the rear of their quarters, but the grounds have for the last forty years or more, been used for agricultural and other purposes and every appearance of its former use has long since disappeared.

The Hospital Barracks were Situated about one mile above the Village, on the bank of Ellicott Creek, and a short distance is their place of burial, covering less than one acre of ground, upon which two hundred or more were buried, some in separate graves, others several in one grave, without coffins, and no possibility of Identifying any, where they have lain for sixty years forgotten by their friends, the busy World and the Government they sacrificed their lives to sustain, And upon the South East corner was also buried in one common grave the English prisoners fatally injured at the affair at Fort Erie.

The Grounds are not fenced except on the Street leading up the Creek, the lands adjoining have bee Continually under cultivation and some think the occupants have encroached upon the grounds of this I cannot say.

Some years ago, there was talk of securing the title for the purpose of having then suitably cared for, but by whom I do not know, nor can I find out by any of the few remaining residents of that time.

Truly Your,
T. A. Hopkins

From the personal collection of Nelson S. Hopkins,
housed at the Buffalo History Museum.
Transcript of Timothy A. Hopkins Papers
Page 9 Corner Number 280

In December after the Declaration of the War of 1812 about 6000 troops were ordered to Williamsville for winter quarters and more or less of them remained there until about the close of Hostilities.

The main body were located on the north side of Main Street and their barracks built of logs extended from the Spring that the Brewery has been built over up nearly to the Creek.

A short distance from the road the Artillery had their quarters between Main Street and the Spring since called Lehn's Spring.

The Hospital Barracks were situated about a mile above the Village on the bank of Ellicott Creek and build of logs on both sides of the road. About a quarter of a mile to their Burial grounds a half or three fourths of an acre, and now with little or no possibility of identifying any of them, on the south east corner of the lot the prisoners that died of the injuries they received at the blowing up of Fort Erie were buried.

At the time the troops were ordered to Williamsville Jonas Williams was absent being a member of the legislature then in session at Albany and when he returned and found his choice timber cut down to build barracks, which he condemned in no measured terms, but there was no remedy except by application to the government.

During the War confusion in every direction, which has been often told and enlarged upon Mr. Williams had build another Sawmill on the west side, and in 1811 built a gristmill having one run of stone driven by a 12 foot overshot water wheel which he got to running by Christmas of that year during that Winter he put in a .....

4
THE CANNON OF THE OLD GARRISON CEMETERY

There have been two different cannon stationed at the Old Garrison Cemetery the first was a Parrott Gun donated by the City of Buffalo, NY in 1898 and the second was an original War of 1812 cannon used by the U.S. Artillery and when captured used by the British which was donated to the cemetery in 1939. Here are their stories.

Dr. Joseph C. Greene of Buffalo was put in charge of the work of enclosing the cemetery and placing the first cannon therein.

History of the Parrott Gun

The Parrott Gun

The first cannon was donated by the City of Buffalo on November 1, 1898, Mayor Diehl of the City of Buffalo, presented a resolution giving permission to the Buffalo Historical Society to place one of the Parrot guns new in the parks at the cemetery at Williamsville, to mark the graves of soldiers of the war of 1812.

Commissioners Day and Conners opposed the granting of the request.

Commissioner Harding said that while he had formerly opposed the request, he was now prepared to vote for it, because he was satisfied that the right of transferring the guns was consistent with a ruling of the war department.

Mayor Diehl said that he had made enquiry. and found no objection among the people who are most vitally interested in historical matters.

Commissioner Sweeney said he had no doubt but 50 members of the 13th regiment, "our own 13th," were buried at Williamsville.

Mayor Diehl's resolution was carried by a vote of 10 to 6.

Secures the Cannon.

The Buffalo Historical Society renewed its request to have a Parrot cannon donated to it from the park system to mark the graves of the soldiers of the War of 1812, buried in the Williamsville Cemetery at the Eleven Mile Creek.

Mayor Diehl moved to grant the request.

Commissioner Day strenuously opposed the granting of the request. He pointed out that there were eight of the Parrot guns at Lafayette Square, six at Soldiers' Place, and nine at the Front. All these guns had been arranged in symetrical order, and to take one away from any place would necessarily create a gap which would force the department to rearrange all the guns. Besides this, the placing of a Parrot gun over the graves of soldiers of the War of 1812 would signify nothing. The Parrot cannon was a modern gun and had not even a remote connection with the War of 1812. Such a cannon erected at Williamsville would be a monument to nothing but the bad taste of the Historical Society. "If each member of the Historical Society," said Mr. Day, 'will pay $10, as I will, then an appropriate monument might be erected over these soldiers' graves. But the idea of erecting a Parrot gun, which is a modern construction, over these soldiers' graves would be useless and inappropriate."

The Mayor and Commissioner Sweeney spoke in favor of granting the request.

On a vote being taken the motion by the Mayor was carried and the Historical Society will get the gun.

The Buffalo Review Nov 02, 1898

On November 8, 1898 the cannon was placed in the center of the cemetery and was inscribed "In memory of the dead soldiers". The Buffalo Historical Society would hold a dedications ceremony the following summer.

TO SOLDIERS OF 1812.

CANNON HAS BEEN PLACED IN THE OLD CEMETERY NEAR WILLIAMSVILLE.

At noon yesterday the Buffalo Historical Society placed a cannon in the old burying ground near Williamsville, to mark the last resting place of American and British soldiers who died in the War of 1812 and were buried together near the spot where the old barracks stood. There were no formal exercises. President Andrew Langdon, Dr. Joseph C. Greene and James Sweeney, representing the Buffalo Historical Society, A. L. Rinewalt of the Amherst Bee, chairman of the Williamsville citizens' committee, and many of the townspeople gathered for the event.

The cannon was placed in the center of a half-acre plat, cleared and inclosed by an ornamental iron fence, in a path eight feet wide, trained toward the arch over the gateway. The cannon is a Parrott gun, taken from the north end of the circle at The Front and is suitably inscribed to the memory of the dead soldiers. Over the gateway is an inscription, "Soldiers of the War of 1812," the letters being cut from galvanized iron.

The Buffalo Historical Society now holds title to the lot by a deed from the heirs of the former owners, and intends to dedicate the cemetery, with appropriate exercises, early next summer. The people of Williamsville will beautify the grounds by planting maple trees.

Buffalo Courier Express Nov 09, 1898

On August 25th, 1942, the original cannon (acquired in 1898) was presented to the Bethlehem Steel Corporation to be used as scrap metal for the war effort.

History of the War of 1812 Cannon

War of 1812 Cannon

The cannon was used by the Americans during Brown's campaign down the Niagara in1814 and was captured by the British. It was employed by the British 3rd Battery during the siege of Fort Erie. After the war it was buried muzzle down and the cascabel used for many years as a hitching post in front of Doctor Douglas' residence in the Village of Fort Erie. In 1936 it was acquired by the Frontier Post No. 71 Canadian Legion of the British Empire Service League, having its headquarters at Fort Erie, Ontario. At that time someone noticed that it bore the ordnance mark of the United States. Once that was discovered in accordance with the spirt of amity which exists between Canada and the United States, the Canadian Legion presented the field piece to the Buffalo Historical Society upon the condition that it be placed in the cemetery as a memorial to the soldiers of both countries. Soon after this generous gift was made, the Buffalo Historical Society procured a blueprint, showing how such guns were mounted during the War of 1812 and they had the gun carriage constructed in exact accordance with such plans. The field piece now appears precisely as it did when in service during the war.

Dedication services were held on Memorial Day, May 30, 1939 under the auspices of the Buffalo Historical Society, acting jointly with the Frontier Post No. 71, Canadian Legion of British Empire Service League, the George F. Lamm Post No. 622 of the American Legion, and officers of the Village of Williamsville and neighboring municipalities.

War of 1812 Cannon Found Here, Ownership Puzzles

Historical Society Director Reveals Prized Field Piece Is Parked Before Buffalo Building

Mystery surrounding whereabouts of 126-year-old cannon, a prized relic from the War of 1812, was cleared up yesterday by Robert W. Bingham, director of Buffalo's Historical Society.

But the question of ownership still was puzzling historians of the United States and Canada.

In Fort Erie, S. C. Wade of Kingston, who represents Canada's Dept. of Antiquary at Ottawa, is searching for the field piece and seeks its return to Fort Erie for a museum in the old fort.

He arrived in Fort Erie armed with newspaper clippings describing the cannon.

Found in City

Meanwhile, in Buffalo, Mr. Bingham cleared up the mystery of the cannon, which, it was reported, Mr. Wade was having trouble locating.

"Why it's out in front of the Historical Building here, facing Pake Lake," Mr. Bingham said. "It's ours."

Back in Fort Erie, Tobias F. Forestell, past president of Post 71 of the Canadian Legion, explained how the cannon came to be in Buffalo.

Scoffs at Demand

"I was president of Post 71 last year when this cannon was donated to Buffalo," he said. "I had full authority for my actions as the cannon was the property of the Canadian Legion at Fort Erie, having been given to us, without any strings attached, by the late Dr. William Douglas, whose property it was for many years.

"I have received no official demand from any department in Ottawa for the return of the cannon and I certainly would pay no serious attention to one if it were made."

For years the field piece, taken from its American owners by British forces in the War of 1812, was used by Dr. Douglas as a hitching post.

Ottawa,, it was said in Fort Erie, is anxious for the cannon's return to use as part of the restored ruins of Old Fort Erie. The $40,000 project is planned for next spring.

The cannon is being prepared here for a Memorial Day service in a cemetery for veterans of the War of 1812 in Williamsville, Mr. Bingham said.

Ceremony Slated

"We've had it mounted," the director said, in announcing Canadian and American veterans would be invited to attend the ceremonies at Williamsville, where the cannon will be placed.

The gun, having a 10-foot length overall, was brought to Buffalo for the parade at the Veterans of Foreign Wars convention. It was given as a gesture of international goodwill by Post 71, it was said.

"It's a regulation American field piece captured by the British in warfare in and around Fort Erie," Mr. Bingham said in describing the object of the Canadian official's search.

The Buffalo Times Feb 06, 1938

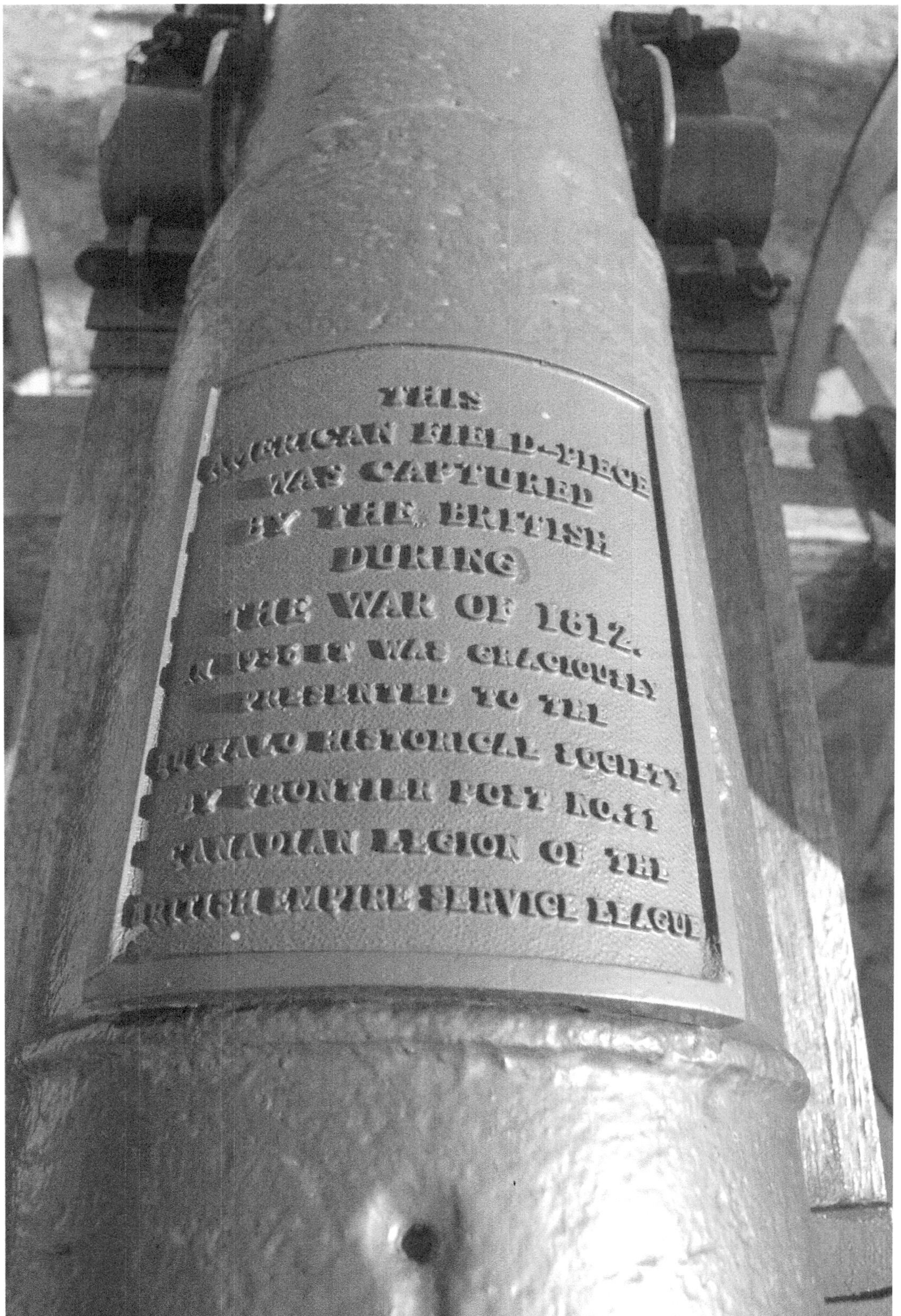

Dedication plaque mounted on War of 1812 cannon.

5
PHOTOGRAPHY OF THE OLD GARRISON CEMETERY

As you can see from this earliest known photograph of the graveyard, the five sugar maple trees on the eastern border of the graveyard as described in the historical account. Possibly One of the Oldest Known Photographs of War of 1812 Cemetery

Most likely 1898 or 1899 after the estates of John Bliss and John R. Evans deeded the cemetery to the Buffalo Historical Society.

Possibly One of the Oldest Known Photographs of War of 1812 Cemetery, prior to gate installation.

Photograph looking east 1898-1899

OLD GARRISON BURYING GROUND.

Photograph looking south

Parrott Gun photograph prior to 1942

Photograph from the 1950's

Photograph looking west May 30, 1939

Photograph looking east date unknown

NYS Education Department Historical Marker installed 1939

William C. Pomeroy Foundation Marker Installed 2021

Main Entrance

. . .

Burial Mound

Burial Mound

Burial Mound

Burial Mound

Facing South East

Facing North West

WAR OF 1812 - 150 YRS. - 1962
DEDICATED, JUNE 17TH. 1962,
TO THE VALOR OF THE AMERICAN
AND BRITISH SOLDIERS, AND
THE PEACE WHICH HAS
LONG ENDURED.
V. F. W. POST 2429
COMMANDER
CHARLES J. WEBER

. . .

Plaque on bolder

DEDICATION OF THE WAR OF 1812 CEMETERY
AT WILLIAMSVILLE, N. Y., MAY 30, 1939. THE HON. JAMES WADSWORTH SPEAKING

6
CEMETERY LAYOUT

As described in the Buffalo Cemeteries: An Account of the Burial-places of Buffalo, from the Earliest Times, read before the Buffalo Historical Society, February 4, 1879:

On the border of two sides stand ten sugar maple trees, from one and one-fourth to two feet in diameter: all but one in a fine, healthy condition. I have no doubt they were planted there by the comrades of those whose bones now occupy this ground. The tenth tree is in a state of decay, and like many of the human race, is beginning to die at the top.

I believe five sugar maple trees on each side which are shown on the left of the photo.

After carefully looking through the archives for maps and aerial photographs of the area in which the burial ground stood, I could compile a scaled overlay of the area onto Google Earth.

The next image is the 1927 aerial photograph of the burial ground. The cemetery marker on the map signifies the location from the canon at the present time. The three remaining sugar maple trees are visible of the original 10 planted and the black dot at the line is possibly the Parrot Gun installed in 1898 by the Buffalo Historical Society.

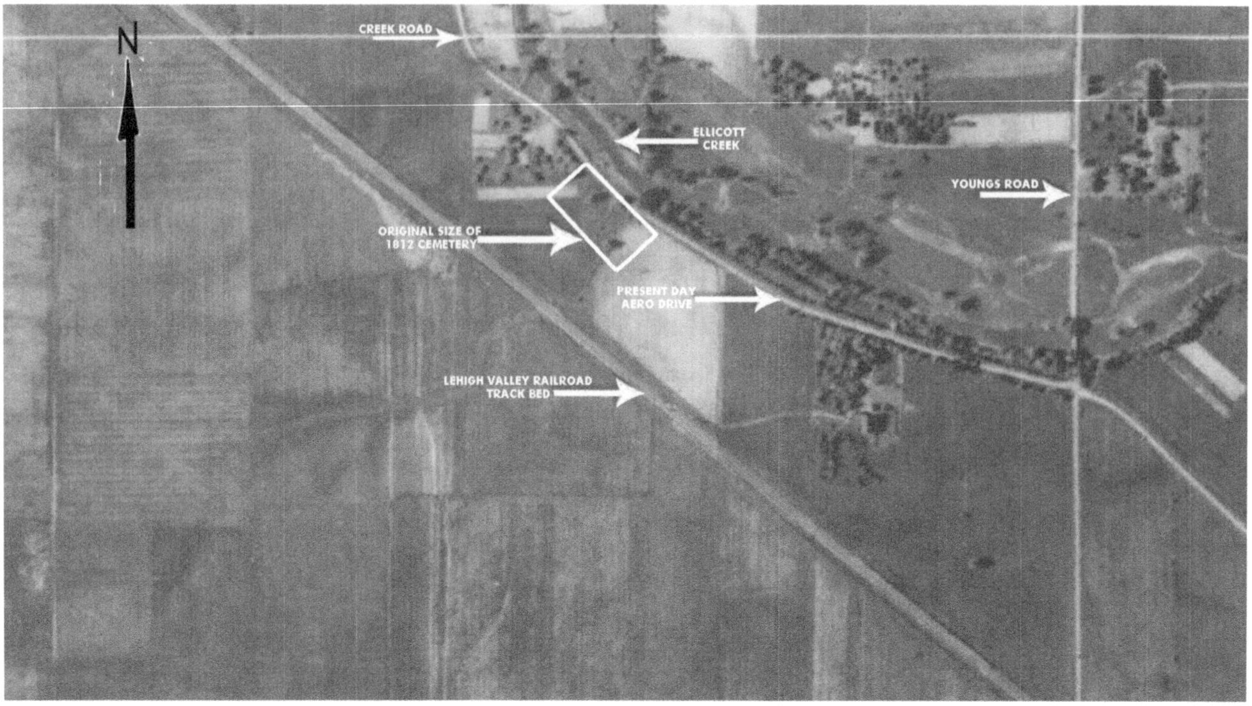

Figure 33: 1927 Aerial Photograph of the burial ground

Figure 34: 1951 Aerial Photograph of the Graveyard

Present-day satellite image of burial ground lines overlaying the boundary of the original graveyard from 1927 aerial photograph.

Figure 35: Lines Represent Current Day burial ground Boundaries

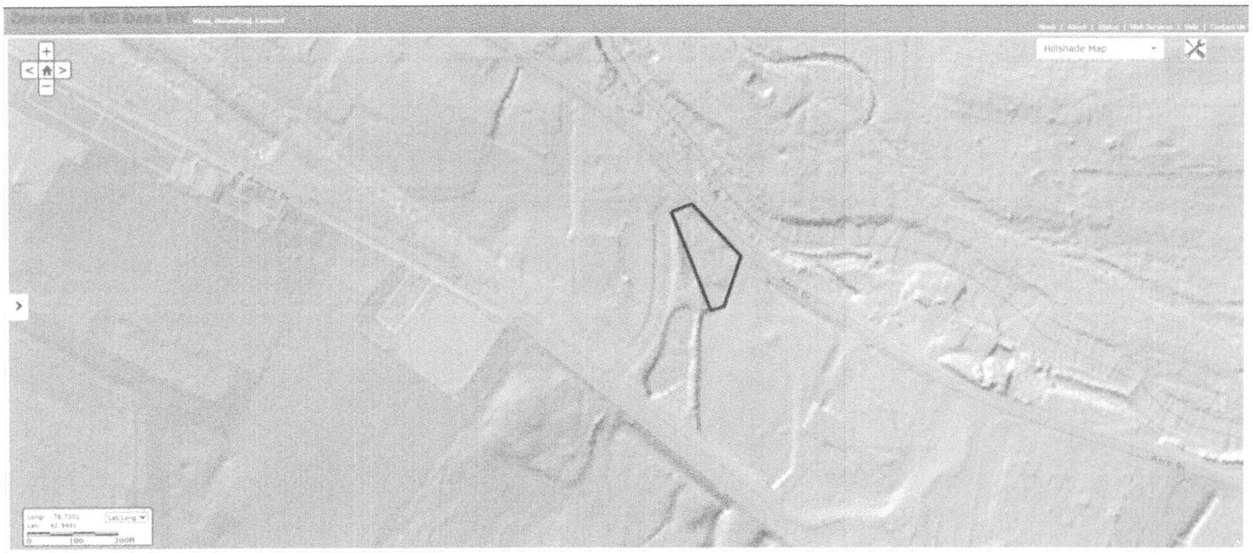

Figure 36: LiDar burial ground Area

In this image the present-day burial ground boundaries are marked and the original burial ground boundaries outlined. As one can see, a good portion of the original burial ground is now a driveway and parking area.

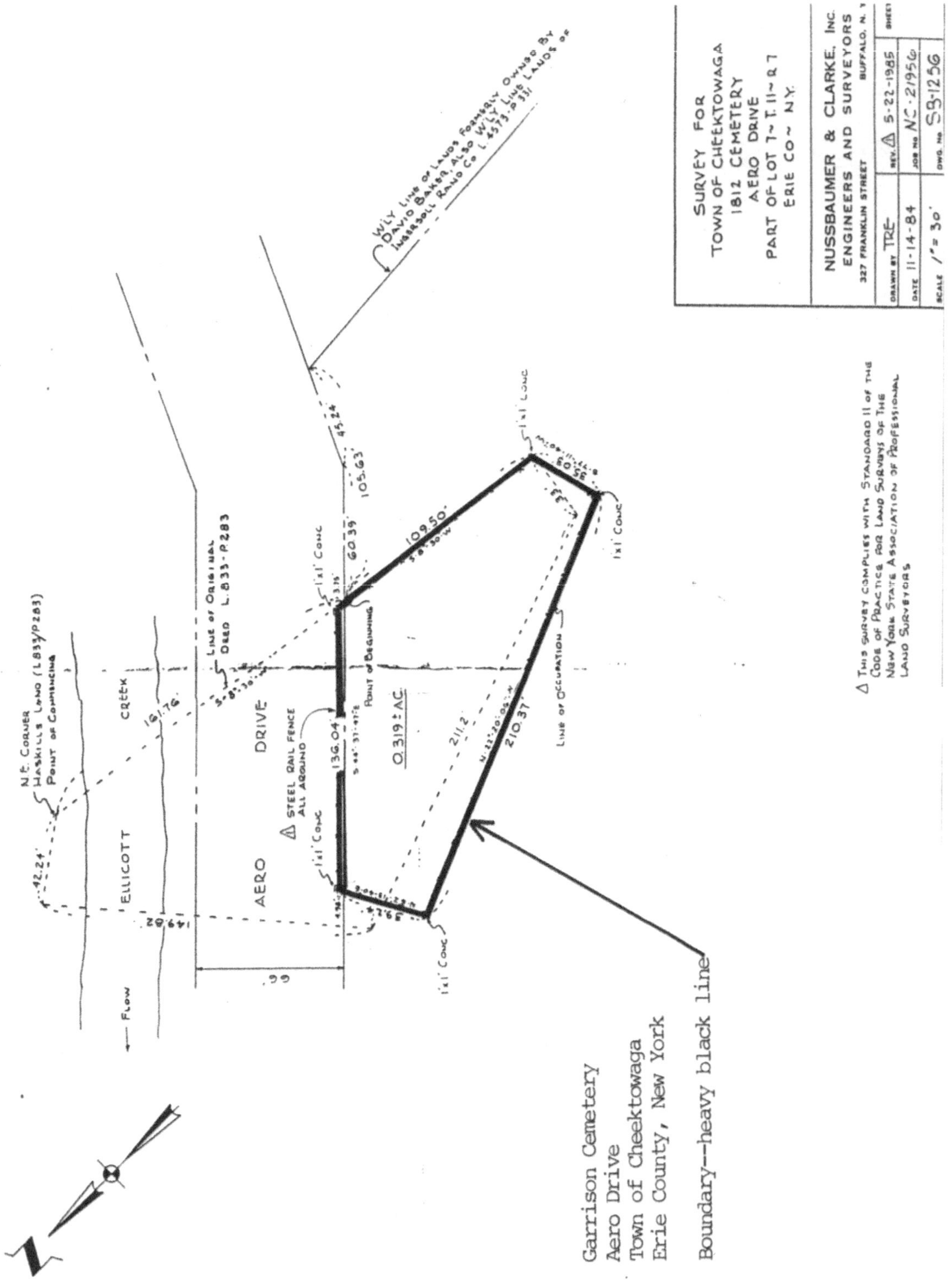
SURVEY FOR
TOWN OF CHEEKTOWAGA
1812 CEMETERY
AERO DRIVE
PART OF LOT 7 ~ T. 11 ~ R 7
ERIE CO ~ N.Y.
NUSSBAUMER & CLARKE, INC.
ENGINEERS AND SURVEYORS
327 FRANKLIN STREET
BUFFALO, N. Y.
DRAWN BY TRE
REV. 5-22-1985
DATE 11-14-84
JOB NO NC-21956
SCALE 1"= 30'
DWG. NO SB-1256
This survey complies with Standard II of the Code of Practice for Land Surveys of the New York State Association of Professional Land Surveyors
ELLICOTT CREEK
AERO DRIVE
FLOW
STEEL RAIL FENCE ALL AROUND
0.319± AC
POINT OF BEGINNING
LINE OF OCCUPATION
LINE OF ORIGINAL DEED L.833-P283
N.E. CORNER HASKILLS LAND (L.833/P283) POINT OF COMMENCING
1x1' CONC
136.04
210.37
211.2
109.50
60.39
105.63
45.24
16.76
42.24
Garrison Cemetery
Aero Drive
Town of Cheektowaga
Erie County, New York
Boundary—heavy black line

. . .

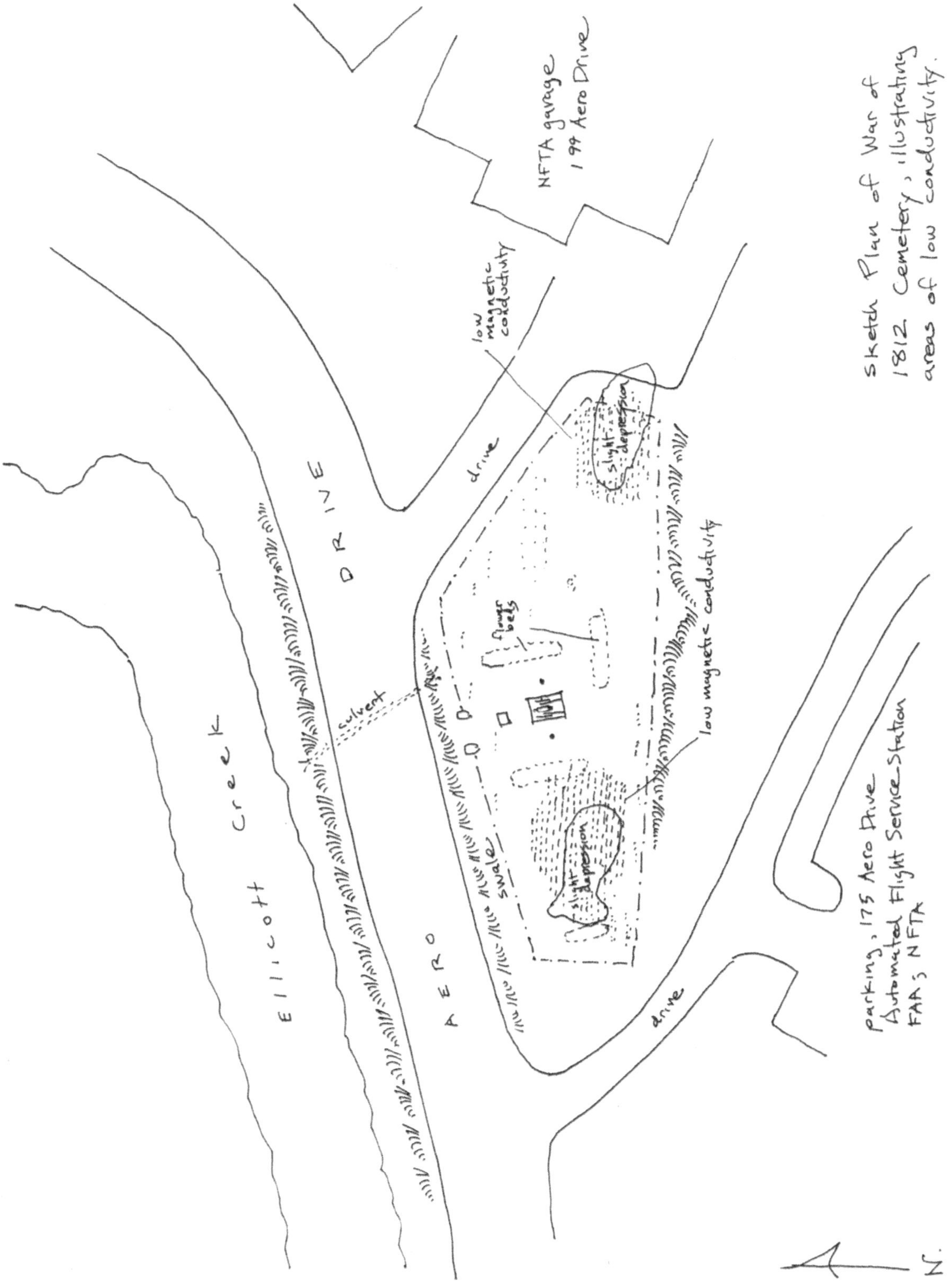
NFTA garage
199 Aero Drive
Sketch Plan of War of 1812 Cemetery, illustrating areas of low conductivity.
low magnetic conductivity
slight depression
drive
Ellicott Creek
AERO DRIVE
low magnetic conductivity
flower beds
culvert
low magnetic conductivity
swale
slight depression
drive
parking, 175 Aero Drive
Automated Flight Service Station
FAA's NFTA
N.

7
LIST OF BURIED U.S. SOLDIERS AT THEWAR OF 1812 GARRISON CEMETERY

In 1813, the local barracks underwent expansion to serve as a hospital with Dr. Mann as the attending surgeon. The facility received and treated over eleven hundred sick and wounded soldiers, including some British prisoners transferred from Lewiston, New York. This event transformed the entire area into a vital hospital base.

Soldiers who succumbed to their injuries during their treatment at this military hospital were laid to rest in a cemetery situated on Aero Drive, then within the boundaries of Amherst. The construction of the New York State Thruway nearly 150 years later necessitated the relocation of the cemetery, which now falls under the jurisdiction of Cheektowaga.

A noteworthy observation is that of the 432 soldiers interred in this cemetery, only twelve hailed from New York State. The remaining soldiers, comprising infantrymen and artillerymen, represented a diverse range of states, including New Hampshire, Vermont, Massachusetts, Connecticut, New Jersey, Pennsylvania, Maryland, Virginia, North Carolina, Ohio, and Kentucky. The database of names were compiled by Cheektowaga Town Historian Maureen E. Gleason and David Przepior.

Name	DoD	Unit	Commanding Officer
Calmon James W	29 Aug 1814	1st US Inf	Capts Symmes & Clemson
Caulker Michael	24 Aug 1814	1st US Inf	Capts Symmes & Clemson
Conker Michael	23 Aug 1814	1st US Inf	Capt John C Symmes
Coulker Michael	24 Aug 1814	1st US Inf	Capt H Stark
Critichlow John	1 Sept 1814	1st US Inf	Capt John C Symmes
Detwiller David	15 Dec 1814	1st US Inf	Capts Hamilton & Owens
Emons John	12 Jan 1815	1st US Inf	Capts Johnson & Symmes
Fegan William	25 Dec 1814	1st US Inf	Capts Stark & Owens
Henderson James	28 Oct 1814	1st US Inf	Capts Symmes & Clemson
Hill Samuel H	1 Jan 1814	1st US Inf	Capt F Hamilton
Kenney Daniel Corporal	11 Sept 1814	1st US Inf	Capt John C Symmes
Kinkaid James	1 Dec 1814	1st US Inf	Capt S Owens
Mc Colman/Mc Calmon James	29 Aug 1814	1st US Inf	Capt John C Symmes
Mc Mullen John	16 Oct 1814	1st US Inf	Capt Simon Owens
Mc Neal Samuel	1 Feb 1815	1st US Inf	Capt John C Symmes
Mellen Jacob	28 Oct 1814	1st US Inf	Capt Hamilton
Miller Jacob	28 Nov 1814	1st US Inf	Capts Clemson & Symmes
Mullen John	18 Oct 1814	1st US Inf	Capt Hamilton
Neal William	19 Jan 1815	1st US Inf	Capt John C Symmes
TenEykck Samuel	11 Sept 1814	1st US Inf	Capt John C Symmes
Morris Joseph	27 Dec 1814	1st US Rifles	Capt Thomas Ramsey
Price Franklin Corporal	11 Dec 1814	1st US Rifles	Capt Thomas Ramsey
Souther Samuel	1/13 Jan 1815	1st US Rifles	Capt Edward Wadsworth
Bridges Daniel	7 Nov 1814	4th US Inf	Capt Josiah Bacon
Chase Benjamin	11 Feb 1815	4th US Inf	Capt John Smith
Colby Nicholas	24 Feb 1815	4th US Inf	Capt Josiah Bacon
Littlefield John	11 Jan 1815	4th US Inf	Capt A Hawkins
Roundy Ralph	10 Feb 1815	4th US Inf	Capt Frederick Conklin
Silver John	25 Jan 1815	4th US Inf	CaptL ewis Peckham
Simonds Sylvester	24 Jan 1815	4th US Inf	Capt Fred Conkling
Sylver John	30 Jan 1815	4th US Inf	Unknown
Trask Nehemiah	17 Feb 1815	4th US Inf	Capt George Gooding
Watts William	18 Nov 1814	4th US Inf	Capt Larabee
Stewart Andrew	16 Dec 1814	4th US Rifles	Capt Joseph Kean
Conlon James	7 Dec 1814	5th US Inf	Capts Dorman & Jamison
Fisher John	3 Nov 1815	5th US Inf	Capt James Dorman
Nagle Christian	17 Apr 1813	5th US Inf	Capt Richard Whartenby
Pryor James	13 Jan 1815	5th US Inf	Capt Leroy Opie
Shaw Thomas	12 Dec 1814	5th US Inf	Capt James Dorman
Andrews Seth	Unknown	6th US Inf	Unknown
Allen Noah Corporal	1 Nov 1814	9th US Inf	Capt Turner Crooker
Chase Elkanah	25 July 1814	9th US Inf	Capt Georgr Bender Jr
Crooker Calvin	16 Oct 1814	9th US Inf	Lt Daniel Chandler
Ditson Joseph	Unknown	9th US Inf	Capt David S Townsend
Eldridge Levi	July/Aug 1814	9th US Inf	Capt Ropes
Elliot James	4 Jan 1814	9th US Inf	Capt Turner Crooker

Fales/Fails Plina	28 Jan 1814	9th US Inf	Lt W L Foster
Hadley Amos	6 Aug 1814	9th US Inf	Capt Thomas J Harrison
Harrington Isaac	28 Dec 1814	9th US Inf	Lt Daniel Chandler
Hart Charles	11 Jan 1814	9th US Inf	Capt D S Townsend
Hart James	20 Nov 1814	9th US Inf	Capt Thomas J Harrison
Heard Timothy	Nov 1813	9th US Inf	Capt Mose Holt
Horsham Joel	Unknown	9th US Inf	Capt Ebenezer Childs
Johnson Jesse	2 Jan 1814	9th US Inf	Capt David S Townsend
Johnson William	12 Sept 1814	9th US Inf	Capt Thomas J Harrison
Keeler Isaac	25 Jan 1815	9th US Inf	Capt Samuel L Allen
Kelph Peter	15 Jan 1814	9th US Inf	Capt J F Norris
Lord Warren	30 Nov 1813	9th US Inf	Capt Ebenezer Childs
Luce Noah	Unknown	9th US Inf	Capt Chester Lyman
Manning Nathaniel Corporal	16 Sep 1814	9th US Inf	Lt Daniel Chandler
Marshall James	22 Jan 1814	9th US Inf	Capt D S Townsend
Mc Lahlan Hugh	Unknown	9th US Inf	Capt Ebenezer Childs
Morey James	Unknown	9th US Inf	Lt Samuel L Allen
Morse Nathaniel	17 Jan 1815	9th US Inf	Capt Ebenezer Childs
Mory James	26 Dec 1814	9th US Inf	Lt Jared Ingersoll
Reed Samuel	27 Oct 1814	9th US Inf	Capt Samuel L Allen
Sealy Ruben	15 Apr 1814	9th US Inf	Capt Norris
Skelton William	20 Nov 1814	9th US Inf	Capt Ebenezer Childs
Whipple William	3 Jun 1815	9th US Inf	Lt Jared Ingersoll
Wyer David	15 Aug 1814	9th US Inf	Capt Thomas J Harrison
Booth Gray	25 Jan 1815	10th US Inf	Capt Joseph Clay
Brooks Eli	30 Jan 1815	10th US Inf	Capt Robert Mitchell
Chesnutt Duke W	15 Feb 1815	10th US Inf	Capt James Wellborn
Danson John	25 Jan 1815	10th US Inf	Capt Joseph Clay
Dawson John	25 Jan 1815	10th US Inf	Capt Joseph Clay
Freet George	17 Feb 1815	10th US Inf	Capt Joseph Clay
Patterson William	15 Jan 1815	10th US Inf	Capt Joseph Clay
Stallings Hardy	1 Feb 1815	10th US Inf	Capt Joseph Clay
Taylor John Sergeant	Unknown	10th US Inf	Capts Vashon & Bailey & Nelson
Taylor Redding	20 Mar 1815	10th US Inf	Capt Robert Mitchell
Tyler Redding	16 Feb 1815	10th US Inf	Capt Robert Mitchell
Wallace John	31 Nov 1814	10th US Inf	Capt Bryant & Nelson
Warick James	22 Dec 1814	10th US Inf	Capt George Vashon
Alvord Elisha	11 Nov 1814	11th US Inf	Capt William S Foster
Atwood Elisha	11 Nov 1814	11th US Inf	Capt William S Foster
Badger Isaac	18 Sept 1814	11th US Inf	Capt William S Foster
Belding Jonathan	25 June 1815	11th US Inf	Lt Isaac Clark
Bradford Timothy	15 Apr 1814	11th US Inf	Capt Valentine Goodrich
Brown Ephraim	13 Aug 1814	11th US Inf	Lt Isaac Clark
Cady Aden	9 Aug 1814	11th US Inf	Capt Samuel H Holley
Campbell James B Colonel	4 Sept 1814	11th US Inf	Unknown
Doolittle Abraham	22 Aug 1814	11th US Inf	Capt B Smead
Ferris Gideon P	24 Aug 1814	11th US Inf	Capt John Bliss

Fry Asa	14 Sept 1814	11th US Inf	Capt William S Foster
Fry Isaac	14 Sept 1814	11th US Inf	Unknown
Goodenow Hezekiah	10 Feb 1815	11th US Inf	Capt William S Foster
Gould John Jr.	30 Oct 1814	11th US Inf	Capt William S Foster
Graves Waine	18 Nov 1814	11th US Inf	Capt William S Foster
Grimes James	19 Sept 1814	11th US Inf	Capt William S Foster
Hall Lyman	26 Oct 1814	11th US Inf	Capt William S Foster
Harris John	26 Aug 1814	11th US Inf	Capt Bliss & Smead
Hildreth Peter	5 Aug 1814	11th US Inf	Capt John Bliss
Hill William	30 Aug 1814	11th US Inf	Capt Valentine R Goodrich
Hull Lyman	1814	11th US Inf	Capt John W Weeks
Lainhard David	25 Aug 1814	11th US Inf	Capt Malachi Corning
Lock James	Unknown	11th US Inf	Capt William S Foster
Lovering Jonathan	1 Nov 1814	11th US Inf	Capt William S Foster
Martin David	30 Aug 1814	11th US Inf	Unknown
Mc Pherson John	Unknown	11th US Inf	Capt John W Weeks
Moulton Richard	25 Sep 1814	11th US Inf	Capt William S Foster
Nichols John	30 Nov 1814	11th US Inf	Capt William S Foster
Parkhurst Thomas	31 Dec 1814	11th US Inf	Capt William S Foster
Peck Henry	27 Sep 1814	11th US Inf	Capt William S Foster
Ross John	22 Sept 1814	11th US Inf	Lt B Smead
Severance Jonathan	12 Nov 1814	11th US Inf	Capt William S Foster
Sovereign Jonathan	1 Nov 1814	11th US Inf	Capt William S Foster
Sprague Eliakiam	20 Oct 1814	11th US Inf	Unknown
Stephens James	30 Jan 1815	11th US Inf	Capt William S Foster
Warren John Corporal	3 Oct 1814	11th US Inf	Capt William S Foster
White John	2 Nov 1814	11th US Inf	Capts Hale & Foster
Whitman John	8 Nov 1814	11th US Inf	Capt William S Foster
Bell Thomas	Jan 1815	12th US Inf	Capt James Paxton
Bowman Adam	28 Oct 1814	12th US Inf	Capt James Paxton
Bowman John	28 Oct 1814	12th US Inf	Capt James Paxton
Grammer Robert	21 Jun 1813	12th US Inf	Capt James Paxton
Helms Abraham	14 May 1814	12th US Inf	Capt James Paxton
James Jacob	13 Mar 1814	12th US Inf	Capt James Paxton
Manly Bay	29 Oct 1814	12th US Inf	Capt Post
Manly Littlebury	29 Oct 1814	12th US Inf	Capt Andrew Madison
Pully Robert	22 Nov 1814	12th US Inf	Capt Willis
Shaw Nathaniel	24 Oct 1814	12th US Inf	Unknown
Turk Benjamin	25 Dec 1813	13th US Inf	Capt M M Dox
Van Sickler Gilbert	25 Dec 1813	13th US Inf	Capt Myndert Dox
Ausant William	10 Jan 1815	14th US Inf	Capt Earle
Austin Benjamin	18 Jan 1815	14th US Inf	Capt Gilder
Brazur Thomas	10 Jun 1815	14th US Inf	Capt Marshall
Browson Oron	Nov 1814	14th US Inf	CaptJ Stanard
Hagan Isaac	8 Feb 1815	14th US Inf	Capt Earle
Lecompte William	10 Jan 1815	14th US Inf	Capt Richard Arell
Peach Issac	8 Feb 1815	14th US Inf	Unknown

Pryor John	22 Nov 1814	14th US Inf	Capt Gilder
Rictor Christian	Nov 1814	14th US Inf	Capt Gilder
Flemming James	3 Nov 1814	15th US Inf	Capt Hayden
Reed Joseph	9 Jan 1815	15th US Inf	Capt H H VanDalsem
Sumners John	24 Oct 1814	15th US Inf	Capt Barton
Talmadge Robert	9 Jun 1815	15th US Inf	Capt H H VanDalsen
Terhune Albert	15 Jan 1815	15th US Inf	Capt Hayden
Barrett Levi	14 Dec 1814	16th US Inf	Capt J Baldy
Bell Joseph	Feb 1814	16th US Inf	Capt J T McElroy
Matson John	10 Dec 1814	16th US Inf	Capts Atkins & Mc Laughlin
Williams john	13 Dec 1814	16th US Inf	Capt Atkins
Alexander Benjamin	10 Jan 1815	17th US Inf	Capts Jacksons & Saunders
Ball Adam	7 Jan 1815	17th US Inf	Capt David Holt
Ballard Linsey	1 Mar 1815	17th US Inf	Capt Trimball
Benjamin Alexander	10 Jan 1815	17th US Inf	Capt H H Hickman
Berkley John	30 May 1815	17th US Inf	Capts Adair & CaptHolt
Brown John	11 Jan 1815	17th US Inf	Capt Adair
Butler Thomas	21 Dec 1814	17th US Inf	Capt David Holt
Campbell Samuel	Feb 1814	17th US Inf	Capt H H Hickman
Canell Castle	21 Dec 1814	17th US Inf	Lt Hackley
Capil Samuel	9 Dec 1814	17th US Inf	Lt Hackley
Capitte Paul	9 Dec 1814	17th US Inf	Capt David Holt
Crawford Michael	7 Dec 1814	17th US Inf	Capt David Holt
Dan Peter	14 Jan 1815	17th US Inf	Capt Hackley
Darnell Samuel S	1 Feb 1815	17th US Inf	Capt Benjamin Sanders
Den Pilee	15 Feb 1815	17th US Inf	Capt H H Hickman
Derr Peter	Feb 1815	17th US Inf	Capt H H Hickman
Duncan George	10 Aug 1814	17th US Inf	Capt John T Chunn
Flemming David	14 Jan 1815	17th US Inf	Capt H H Hickman
Forlorn Meredith	21 Jan 1815	17th US Inf	Capt Mordecai Myers
Fryman Henry	Unknown	17th US Inf	Capt H H Hickman
Hardisty George	11 Feb 1815	17th US Inf	Unknown
Hopkins Isom	15 July 1815	17th US Inf	Capt Scott
Johns John	25 Mar 1815	17th US Inf	Capt David Holt
Kennel Casper	21 Dec 1814	17th US Inf	Capt H H Hickman
Layor John	28 Feb 1815	17th US Inf	Capt William J Adair
Lemmon Benjamin	27 Dec 1814	17th US Inf	Capt David Holt
Marsh William	29 Jan 1815	17th US Inf	Capt H Hickman
Mays John	31 Dec 1814	17th US Inf	Capt John T Chunn
Mc Amy Samuel	15 Jan 1815	17th US Inf	Capt John T Chunn
Mc Combs John	17 June 1814	17th US Inf	Capt Trimbal
Mc Donald Stephen	10 Jan 1815	17th US Inf	Capt David Holt
Moules James	15 Jan 1815	17th US Inf	Unknown
Nation Frederick	21 March 1815	17th US Inf	Capt Henry Crittenden
Norris Beverly	25 Dec 1814	17th US Inf	Capt Benjamin Sanders
Palmer Richard	8 Dec 1814	17th US Inf	Capt H H Hickman
Parmer Richard	8 Dec 1814	17th US Inf	Unknown

Patterson Alexander Lt	Unknown	17th US Inf	Unknown
Robbins Hiram	13 Jan 1815	17th US Inf	Unknown
Roberts Hiram	Jan 1815	17th US Inf	Capt H H Hickman
Sayor John	1 Mar 1815	17th US Inf	Capt Adair
Sidebottom Wilson	18 Mar 1815	17th US Inf	Capt Adair
Smith Joel	24 Dec 1814	17th US Inf	Lt Holt
Sprague Ephraim	25 Jan 1815	17th US Inf	Capt Hickman
Stewart Alex	7 Dec 1814	17th US Inf	Capt B W Sanders
Bear Adam	1 Dec 1814	19th US Inf	Capt Carey A Trimble
Bledsoe Jacob	15 Oct 1814	19th US Inf	Capt William McDonald
Denman Samuel	30 Dec 1814	19th US Inf	Capt Carey A Trimble
Duncan Samuel	6 Jan 1814	19th US Inf	Capt William McDonald
Flemming Patrick	5 Nov 1814	19th US Inf	Capt John T Chunn
Gluffan Thomas	22 Oct 1815	19th US Inf	Capt William McDonald
Griffin Thomas	22 Oct 1814	19th US Inf	Capt William McDonald
Mc Clintock John	13 Sep 1814	19th US Inf	Lt Riddles
Moles James	26 Jan 1815	19th US Inf	Capt H H Hickman
Stansbury Christian	Jan 1815	19th US Inf	Capt Carey A Trimble
Topping Zephar	27 Sep 1814	19th US Inf	Capts McDonald & Trimble
Davis Willis	Sept/Oct 1813	20th US Inf	Capt J Macrae
Franklin Joseph	30 Dec 1813	20th US Inf	Capt John Stanard
Mills William	1 Apr 1813	20th US Inf	Lt T M Randolph
Ames Daniel	1 Jan 1814	21st US Inf	Capt John T Chunn
Baker Amos	4 Oct 1814	21st US Inf	Capt Sullivan Burbank
Bidman/Bidmore Peter	30 Oct 1814	21st US Inf	Cap tWillis Foulk
Grouard James	5 May 1815	21st US Inf	Unknown
Hibbert Jeremiah	15 Feb 1815	21st US Inf	Capt James Pratt
Hutchings William	25 Feb 1815	21st US Inf	Capt Lemuel Bradford
Ireland Joel	Unknown	21st US Inf	Capt Lemuel Bradford
Lincoln Peris	26 Sept 1814	21st US Inf	Capt Sullivan Burbank
Littlefield George Sergeant	1 Feb 1815	21st US Inf	Capt Benjamin Ropes
Millhallen Daniel	Unknown	21st US Inf	Capt Henderson
Mudge Charles	19 Aug 1814	21st US Inf	Capt Sullivan Burbank
Packard/Packer George	Oct 1814	21st US Inf	Capt Sullivan Burbank
Pattee Ezekeil	4 Nov 1814	21st US Inf	Capt Morrill Marston
Peckan George	27 Nov 1814	21st US Inf	Capt Benjamin Ropes
Pettingall William	16 July 1816	21st US Inf	Capt Cass
Petty Ezekial	4 Nov 1814	21st US Inf	Capt Morrill Marston
Philbrook Jonathan	18 Dec 1814	21st US Inf	Capt Joseph Treat
Priest Bartholemew	27 Aug 1814	21st US Inf	Capt Bowman
Smith Charles Sergeant	11 Oct 1814	21st US Inf	Capt Morrill Marston
Smith John	10 Oct 1814	21st US Inf	Capt Joseph Treat
Smith John	28 Oct 1814	21st US Inf	Capt William Bowman
Spearin James	Unknown	21st US Inf	Capt Lemuel Bradford
Stewart John	24 Apr 1815	21st US Inf	Unknown
Stewart Lemuel	12 Dec 1814	21st US Inf	Capt Joseph Treat
Tenney Samuel	18 Jan 1815	21st US Inf	Capt Morrill Marston

Thomson Curtis	11 Nov 1813	21st US Inf	Capt Joseph Treat
Wheddon Samuel	29 Jan 1815	21st US Inf	Capt Ira Drew
Whidden Timothy	25 Feb 1815	21st US Inf	Capt Lemuel Bradford
Witham Levi	1 Dec 1814	21st US Inf	Capt Marston
Clark Thomas	17 Oct 1814	22nd U S Inf	CaptJ oseph Henderson
Armstrong John Lieutenant	28 Aug 1814	22nd US Inf	Capt Willis Foulk
Bell James	30 Oct 1814	22nd US Inf	Unknown
Billman Peter	30 Oct 1814	22nd US Inf	Capt James Henderson
Bowhill John	Unknown	22nd US Inf	Capt William Morrow
Brady Hugh Colonel	Unknown	22nd US Inf	Unknown
Byron Edward	25 Aug 1814	22nd US Inf	Capt Jos Henderson
Caldwell Andrew	30 Oct 1814	22nd US Inf	Col Hugh Brady
Cannon Martin	10 Sept 1814	22nd US Inf	Colonel Hugh Brady
Carman Martin	10 Sept 1814	22nd US Inf	Capt Morris or Henderson
Cochran Willam	1 Oct 1814	22nd US Inf	Unknown
Coughman William	Oct 1814	22nd US Inf	Capt Thomas Lawrence
Crouse John Corporal	20 Sept 1814	22nd US Inf	Capt Anderson
Cummings James	11 Sept 1814	22nd US Inf	Lt John R Guy & Willis Foulk
Gladin William	20 Feb 1815	22nd US Inf	Capt Willis Foulk
Griffy John	10 Dec 1814	22nd US Inf	Lt Richards
Guthrie James	5 Dec 1814	22nd US Inf	Capt Sampson L King
Hammon William	11 Sept 1814	22nd US Inf	Capt Sampson L King
Harble Anthony	1 Oct 1814	22nd US Inf	Capt Willis Foulk
Harman George	11/17 Sept 1814	22nd US Inf	Capt Joseph Henderson
Harmon William	17 Sept 1814	22nd US Inf	Capt Joseph Henderson
Hawkesbury Adam	2 Sept 1814	22nd US Inf	Capts Henderson & Cormack
Hickman William Sergeant	1814	22nd US Inf	Capt John Foster
Hock Grunby Sergeant	14 Oct 1814	22nd US Inf	Capt Willis Foulk
Hook Greenbury Sergeant	14 Oct 1814	22nd US Inf	Capt Willis Foulk
Hoover Jacob Corporal	26 Oct 1814	22nd US Inf	Capt Joseph Henderson
Izard Jonathan	Oct 1814	22nd US Inf	CaptJohn R Guy
Lowre John	19 Sep 1814	22nd US Inf	Lt John Brady
Martial/Marshall George	17 Oct 1814	22nd US Inf	Unknown
McDay Edward	27 Oct 1814	22nd US Inf	Capt Joseph Henderson
McIlhenny William	11 Jan 1815	22nd US Inf	Capt Willis Foulk
Noggle Peter	19 Aug 1814	22nd US Inf	Capt Sampson L King
Prier John	23 Oct 1814	22nd US Inf	Capt Sampson L King
Ramby/Ramsey Nathan	2 Nov 1814	22nd US Inf	Capts Morrow & Henderson
Reddick George	19 Feb 1815	22nd US Inf	Capt William Morrow
Rediur/Redier Jacob	31 Oct 1814	22nd US Inf	Capt Joseph Henderson
Ricker John	5 Nov 1814	22nd US Inf	Capt Joseph Henderson
Rothwell Jacob	Unknown	22nd US Inf	Capt D Espy
Savage Samuel	30 Oct 1814	22nd US Inf	Capt William Morrow
Smith William	16 Sep 1814	22nd US Inf	Capt Joseph Henderson
Stout David	18 Sept 1814	22nd US Inf	Capts Morris & Henderson & King
White George	13 Sep 1814	22nd US Inf	Capt Thomas Lawrence
Wright George	15 Sep 1814	22nd US Inf	Capt Sampson L King

Moore Alexander Corporal	23 Aug 1814	22nd US Inf	Capt Willis Foulk
Moore Charles	Dec 1814	22nd US Inf	Capt John Pentland
Guy John	7 Sept 1814	23nd US Inf	Capt Ingersoll
Bennett Paul	21 Sept 1814	23rd US Inf	Unknown
Brown Walter	4 Jan 1815	23rd US Inf	Capt Richard Goodell
Burrell William	31 Oct 1814	23rd US Inf	Capt F Brown
Cooper Lewis	Unknown	23rd US Inf	Capt Azariah W Odell
Gay John	7 Sept 1814	23rd US Inf	Lt Justus Ingersoll
Holdin Rufus	17 Sept 1814	23rd US Inf	Capt Richard Goodell
Johnson Jonathan	19 Sept 1814	23rd US Inf	Lt Justus Ingersoll
Lawrence Benjamin Corporal	29 Dec 1814	23rd US Inf	Lt Castine
Mc Donald David	13 Sep 1814	23rd US Inf	Capt Justus Ingersoll
Miller Solomon	4 Oct 1814	23rd US Inf	Capt Whiting
O' Riley George	25 Sep 1814	23rd US Inf	Capt Richard Goodell
O' Riley John	6 Sep 1814	23rd US Inf	Capt Richard Goodell
Quimby James	27 Oct 1814	23rd US Inf	Capt Abraham Hull
Ripley	5 Nov 1814	23rd US Inf	Capt Simeon D Wattles
Salisbury Simon	11 Sept 1814	23rd US Inf	Capt Samuel Tappan
Thayer Nathaniel	23 Aug 1814	23rd US Inf	Capt Justus Ingersoll
Weeks Samuel	27 Sep 1814	23rd US Inf	Capt Samuel Tappan
Wiley Henry	13 Nov 1814	23rd US Inf	Lt Samuel Spafford
Keeler John	13 Jan 1815	24th US Inf	Capt John T Chunn
Brace Stephen Jr	7 Jan 1815	25th US Inf	Lt Thomas Seymour
Braman Daniel	31 Oct 1814	25th US Inf	Capt Daniel Ketchum
Champlin George	30 Aug 1814	25th US Inf	Capt Daniel Ketchum
Clark Shaderack	1 Nov 1814	25th US Inf	Lt Thomas Seymour
Cook Aaron	23 Dec 1814	25th US Inf	Capt Daniel Ketchum
Corning David	31 Aug 1814	25th US Inf	Capt Daniel Ketchum
Cottar Joseph	15 Feb 1815	25th US Inf	Lt Thomas Seymour
Hard Hiram/Niram	28 Aug 1814	25th US Inf	Unknown
Haskins Levi	7 Oct 1814	25th US Inf	Capt Daniel Ketchum
Hastings Joseph	19 Sept 1814	25th US Inf	Capt George Howard
Howard Jonathan	7 Nov 1814	25th US Inf	Lt Thomas Seymour
Jennings Noah	11 May 1814	25th US Inf	Capt John B Murdock
Knowlton John	19 Sept 1814	25th US Inf	Capt Seymour
Liswill Thomas	16 Nov 1814	25th US Inf	Capt Thomas Seymour
Lomis Samuel	7 Oct 1814	25th US Inf	Capt Daniel Ketchum
Mansfield Samuel	27 Sep 1814	25th US Inf	Capt Benjamin Watson
Marcellus Cornelius	19 Aug 1814	25th US Inf	Lt Edward White
Miller Asahel Sergeant	5 Oct 1814	25th US Inf	Capt Thomas Seymour
Nelson Silas	26 Oct 1814	25th US Inf	Capt Daniel Ketchum
Potter James Sergeant	22 Aug 1814	25th US Inf	Capt George Howard
Rusco Gamlien	6 Aug 1814	25th US Inf	Lt Thomas Seymour
Schofield Joseph	19 Sept 1814	25th US Inf	Capt Seymour
Scott John	23 Aug 1814	25th US Inf	Capt Daniel Ketchum
Smith Reuben	23 Sept 1814	25th US Inf	Capt Daniel Ketchum
Story Nathan Corporal	10 Nov 1814	25th US Inf	Capt Daniel Ketchum

Stoughton Russel Jr	23 Sept 1814	25th US Inf	Capt Daniel Ketchum
Taylor Elisha	25 Oct 1814	25th US Inf	Capt Daniel Ketchum
Taylor Peter	1 Nov 1814	25th US Inf	Lt Thomas Seymour
Tuttle Charles	6 Aug 1814	25th US Inf	Capt Daniel Ketchum
Waldron Nicholas	25 Feb 1815	25th US Inf	Capt Thomas Seymour
Weed Scridder/Scudder	29 Sep 1814	25th US Inf	Lt Thomas Seymour
Winchel Shaylor	27 Sep 1814	25th US Inf	Capt Daniel Ketchum
Wright Judah	1 Nov 1814	25th US Inf	Capt Daniel Ketchum
Hayette Elisha	15 Oct 1814	27th US Inf	Capt Charles L Cass
Robinson Zadock	22 Dec 1814	Engineers	Capt J W Goodin
Currier Joseph	11 Sept 1814	Sappers	Capt David B Douglas
Matchel/Matchett Robert	26 Dec 1814	1st US Arty	Unknown
Banister Luther	Apr 1815	US Arty	Capt G W Melvin
Bostwick Richmond - Sergeant	2 Sept 1814	US Arty	Capt Alex Williams
Bower Andrew	15 Jan 1815	US Arty	Capt Crary
Bowers Andrew	15 Jan 1815	US Arty	Capt H K Craig
Bowling Simon	17 Feb 1815	US Arty	Capt A C W Fanning
Brickman Joseph	6 Oct 1814	US Arty	Capt Thomas Biddle
Broadwater David	31 Oct 1814	US Arty	Capt Alex J Williams
Calvin William	31 Oct 1814	US Arty	Capt Thomas Biddle
Carter Jacob	14 Sept 1814	US Arty	Capt Thomas Biddle
Colvin William	31 Oct 1814	US Arty	Capt Thomas Biddle
Custer Jacob	Sept 1814	US Arty	Capt Thomas Biddle
Giberson William	Sept 1814	US Arty	Capt Thomas Biddle
Gibson Nathanial	9 Sept 1814	US Arty	Capt John Ritchie
Gibson Thomas Corporal	11 Sept 1814	US Arty	Capt Richie
Girton Jonathan	31 Dec 1814	US Arty	Capt Samuel B Archer
Giverson William	7 Oct 1814	US Arty	Lt Douglas
Goodwin Peter	11 Sept 1814	US Arty	Capt Thomas Biddle
Hickman James	2/3 Jan 1815	US Arty	Capt A J Williams
Jamison Rich	18 Mar 1815	US Arty	Capt Alexander Williams
Morrell Joseph	24 Oct 1814	US Arty	Col Scott & Williams
Paselo Henry	10 Sep 1814	US Arty	Capt Thomas Biddle
Pierce John	23 Aug 1814	US Arty	Capt A J Williams
Pike Artemas	26 Oct 1814	US Arty	Capt Thomas Biddle
Randon Paul	6 Apr 1815	US Arty	Capt Alex J Williams
Stiles Eli	1 Sept 1814	US Arty	Capt A J Williams
Sutter Tobat/Tobal	1 Oct 1814	US Arty	Capt Thomas Biddle
May John F	7 Dec 1814	US Arty	Capt Thomas Ketchum
Morrill Moses	11 Aug 1814	US Atry	Capt Thomas Biddle
Seeley Jeremiah	14 Sept 1814	US Atry	Capt John W Goodins
Skinner Joshua	9 Feb 1815	US Atry	Capt J W Goodin
Vosburgh Russell	16 Mar 1814	US Atry	Capt A W Thorton
Buckman Francis	2 Mar 1814	US Lt Arty	Capt Thornton
Burkans Simon	24 Mar 1814	US Lt Arty	Capt Branch
Foster Luther C	26 Dec1813	US Lt Arty	Capts Branch & Stribling
Baker D	1 Oct 1814	US Lt Dragoons	Capt Haynes

Baldwin Nathaniel	3 Mar 1815	US Lt Dragoons	Capt Haig
Blay Francis	27 Jan 1815	US Lt Dragoons	Capt Samuel D Harris
Lint Lewis	13 Jan 1815	US Lt Dragoons	Capt Large
Holder Daniel	Unknown	US Rifles	Capt William Smyth
Blodgett Clark	13 Sept 1814	NY milita	Capt Moody R Freeman
Hewes William	4 Aug 1814	NY milita	Capts Freeman & Valentine
Paul John	25 Sep 1814	NY milita	Capt Nathaniel F Knapp
Staples Joshua	25 Sept 1814	NY Militia	Capt Joshua Harkin
Graves Lebins	25 Sept 1814	NY or PA Vol	Unknown
Norris Joseph	30 Aug 1814	NY or PA Vol	Capt Brooks
Shells Henry	9 Oct 1814	NY or PA Vol	Unknown
Grey Duncan	25 Sept 1814	NY Vol	Capt Henry Parker
Green Nathanial	9 Sept 1814	NY Vol Cavalry	Capt Haig
Bocher Henry	23 Aug 1814	PA Vol	Capt Samuel Gorden
Brown Alex	16 Sept 1814	PA Vol	Capt James Piper
Cooper Thomas	14 Aug 1814	PA Vol	Capt Andrew Mitchell
Kelley Morgan	30 Aug 1814	PA Vol	Capt Andrew Mitchell
Mawheter William	16 Sep 1814	PA Vol	Capt John Roberts
Woodburn Robert	4 Sep 1814	PA Vol	Capt James Piper
Myers George	12 Sep 1814	PA Vol	Cap tSamuel White
Abbott Daniel L	13 Mar 1814	Unknown	Unknown
Appleby Nathan	16 Oct 1814	Unknown	Capt Foster
Barney John	4 Oct 1814	Unknown	Unknown
Boynton Abel Colonel	2 Sept 1814	Unknown	Unknown
Braden George	21 May 1814	Unknown	Capt Moore
Burrans Semor	24 Mar 1814	Unknown	Capt McIntire
Collin James	30 May 1814	Unknown	Capt Goodridge
Curthburgh	16 Feb 1814	Unknown	Capt McIntire
Elwell Robert	18 Apr 1814	Unknown	Capt Barker
Frick David	4 Oct 1814	Unknown	Capt Cobley
Gale John	7 May 1814	Unknown	Capt Larabee
Griffin Theophiles	14 Sept 1814	Unknown	Capt Hawkins
Hull Simon	July 1814	Unknown	Capts Weeks & Hull
Johnson David	7 Apr 1814	Unknown	Capt VanDolson
Lewis John	2 Mar 1814	Unknown	Capt Thorington
Lewis Lewis	24 May 1814	Unknown	Capt Mutchenburgh
Mann John	22 Mar 1814	Unknown	Capt Barker
Martin Thomas	29 Sep 1814	Unknown	Capt Spencer
Mc Nally John	25 May 1814	Unknown	Capt Muhlenberg
Morton Obed Sergeant	9 Dec 1814	Unknown	Capt Launders
Nichols James	15 Oct 1814	Unknown	Unknown
Perkans James	6 Oct 1814	Unknown	Capt Eastman
Pummons John	24 Oct 1814	Unknown	Capt Bartons
Smith Elias	30 Mar 1814	Unknown	Capt Fuller
Smith Lyman Corporal	23 Sept 1814	Unknown	Capt Douglas
Stickney Abijah	30 Sept 1814	Unknown	Capt Mumford
Swany Edward	23 May 1814	Unknown	Capt Branches

Thompson Thomas	8 Apr 1814	Unknown	Capt McCoon
Thrahal Jacob	1 Mar 1814	Unknown	Unknown
Tibbitts John	5 Sept 1814	Unknown	Unknown
Tibbs Richard	Unknown	Unknown	Capt Moore
Toby Thomas	27 Feb 1814	Unknown	Capt Ball
Twiss Samuel	30 Sept 1814	Unknown	Capt Stridin
Wells Stephen	5 Oct 1814	Unknown	Capt White
Wheeton William	19 Oct 1814	Unknown	Capt Hawkins
White John	13 Mar 1814	Unknown	Capt Thorington
Young Nathaniel	16 Nov 1814	Unknown	Capt Kinney

Also, many British soldiers known only to God.

8
BURIAL GROUND TIMELINE

According to records from the War Department, the Garrison Burying Ground served as a mass burial site for American soldiers beginning in the spring of 1813. The initial interments were those who died in the hospital huts established following that spring, followed by additional burials from the general hospital that became operational in the summer of 1814. Although the surviving documentation is incomplete, it is believed that many more American soldiers were laid to rest in this mass grave.

In addition to American casualties, an undetermined number of British soldiers—captured during the Niagara Campaign of 1814—were also interred at this location.

The earliest land purchase for the military cemetery, situated in Lot 7, Township 11, Range 7, was part of a larger transaction by the Holland Land Company. On October 8, 1808, a total of 202 acres encompassing Lots 6 and 7 were sold to Frank Ludlow. According to Holland Company Records, Ludlow maintained ownership until January 11, 1816, when the land was reclaimed due to his failure to make the required payments. (Holland Company Records).

Then, on 11 January 1816, Wilhem Willink and colleagues (Holland Land Co.), sold property containing the cemetery to Benjamin Barton. (Liber 3 page 133)

17 October 1821, Barton sold property containing the cemetery to Thomas Coe. (Liber 7 page 207)

4 July 1851 William Coe sold property containing the cemetery to John Haskell. (Liber 172, page 459)

6 August 1851 John Haskell, sold the cemetery property only (parcel A) to John Bliss and John R. Evans (Liber 123 page 19) as, "The ground should not be disturbed or encroached upon."

12 July 1898, the estates of John Bliss and John R. Evans deeded the cemetery (Parcel A) to the Buffalo Historical Society. (Liber 833 page 283)

9 November 1898, the Buffalo Historical Society placed a cannon on the burial site. The cannon was a Parrott Gun taken from the north end of the circle at The Front and was suitably inscribed to the memory of the dead soldiers. They installed an arched gateway with the letters Soldiers of the War of 1812 cut from galvanized iron.

By 1918, the old cemetery had fallen into disrepair. The Buffalo Historical Society responded by making improvements again to the cemetery.

In 1932 after much research, the Buffalo Historical Society located the hospital register which was in the possession of the federal government and retrieved the names of over 200 American soldiers who died in the General Military Hospital. It began on 1 August 1814 and the remains were buried at the cemetery. They did not find the names of the soldiers buried before that date until 2011 (see entry for 2011).

30 May 1933, a Memorial Day service was held to rededicate the cemetery in honor of the soldiers now identified and buried in the cemetery. The Buffalo Historical Society made improvements to the burial site, removing the arch and adding boulders with plaques inscribed as follows:

(1) This property is owned and maintained by the Buffalo Historical Society

(2) This tablet is erected by the Buffalo Historical Society in honor of the American and British soldiers of the War of 1812, who died in the U.S. Military Hospital at Williamsville, New York and maintained by the Buffalo Historical Society

In 1936, the Frontier Post #71, Canadian Legion of the British Empire Service League in Fort Erie Ontario, presented a cannon captured from the Americans in the War of 1812.

30 May 1939, a ceremony was held by the Buffalo Historical Society to dedicate the new cannon. A carriage had been built for the cannon which was placed in the middle of the cemetery along with two flag poles erected for American and British flags.

18 June 1939, the Town of Cheektowaga sponsored a dedication at the cemetery on the occasion of the town's centennial celebration.

25 August 1942, the original cannon (acquired in 1898), was presented to the Bethlehem Steel Corporation to be used as scrap metal for the war.

During the 1940s and 50s, Frank Roman of Ellicott Creek Road cared for the cemetery on behalf of the Buffalo Historical Society. In addition, the George F. Lamm Post remembered the veterans by placing flags at the graves on occasion.

17 June 1962, Cheektowaga Post #2429 VFW along with the Buffalo Historical Society rededicated the cemetery for the 150th anniversary of the War of 1812. American and Canadian units took part, and they added a monument with the inscription:

War of 1812 - 150 Years - 1962 Dedicated June 17, 1962, to the valor of the American and British Soldiers, and the peace that has long endured

During the 1960s, Cheektowaga Post #2429 VFW took responsibility for the upkeep of the cemetery.

5 October 1975, Cheektowaga celebrated the U.S. Bicentennial with representatives from both the United States and Canada taking part in ceremonies at the burial site.

Cheektowaga acquired the 1812 Cemetery

In October, 1984 Cheektowaga took the first steps in acquiring the cemetery with a firm hired to do the search and survey.

On 28 January, 1985 the Cheektowaga town Board approved a resolution to acquire the cemetery property. The town agrees to "maintain the War of 1812 Cemetery and to forever dedicate such premises as a cemetery and resting for the soldiers who fought in the War of 1812"

20 March, 1985 a quit-claim deed is signed conveying the cemetery (parcel A) from the Buffalo and Erie County Historical Society to the Town of Cheektowaga. (Liber 9444 page 450)

On 21 March, 1985 the Cheektowaga Town Board approves a resolution transferring ownership of the cannon to the town. "The gift was made subject to the condition that the cannon be perpetually maintained and be forever dedicated as a memorial to the British and American soldiers who fought during the War of 1812".

23 May 1985 a quit claim deed is signed conveying additional cemetery property (parcel B) from Pricilla Brown to the Town of Cheektowaga. (Liber 9450 page 680)

On 9 June 1985 a dedication is held to commemorate the conveyance of the cemetery to the Town of Cheektowaga. The re dedication included representatives from both the United States and Canada. A letter was received from Queen Elizabeth II commending the town.

In 2002 The cemetery is listed on the National Register of Historic Places.

2011 Jack Bilow publishes “A War of 1812 Death Register” after searching the National

Archives for six months. The register adds the names of the soldiers buried at Williamsville prior to 1814.

2021 Cheektowaga Historical Association received a grant from the William G. Pomeroy Foundation to install historical marker.

9

NATIONAL REGISTER OF HISTORIC PLACES

In 2002 an application for the Garrison cemetery to be place on the National and State Historic Register the following page are the completed application, supporting documents and letters of support.

NPS Form 10-900
(Oct. 1990)

OMB No. 10024-0018

United States Department of the Interior
National Park Service

RECEIVED 2280

AUG 27 2002

NAT. REGISTER OF HISTORIC PLACES
NATIONAL PARK SERVICE

1113

National Register of Historic Places Registration Form

This form is for use in nominating or requesting determinations for individual properties and districts. See instructions in *How to Complete the National Register of Historic Places Registration Form* (National Register Bulletin 16A). Complete each item by marking "x" in the appropriate box or by entering the information requested. If an item does not apply to the property being documented, enter "N/A" for "not applicable." For functions, architectural classification, materials, and areas of significance, enter only categories and subcategories from the instructions. Place additional entries and narrative items on continuation sheets (NPS Form 10-900a). Use a typewriter, word processor, or computer, to complete all items.

1. Name of Property

historic name Garrison Cemetery

other name/site number War of 1812 Cemetery

2. Location

street & town Aero Drive ☐ not for publication

city or town Cheektowaga ☐ vicinity

state New York code NY county Erie County code 029 zip code 14227

3. State/Federal Agency Certification

As the designated authority under the National Historic Preservation Act, as amended, I hereby certify that this ☒ nomination ☐ request for determination of eligibility meets the documentation standards for registering properties in the National Register of Historic Places and meets the procedural and professional requirements set forth in 36 CFR Part 60. In my opinion, the property ☒ meets ☐ does not meet the National Register criteria. I recommend that this property be considered significant ☐ nationally ☐ statewide ☒ locally. (☐ See continuation sheet for additional comments.)

[signature] SHPO 8/21/02
Signature of certifying official/Title Date

State of Federal agency and bureau

In my opinion, the property ☐ meets ☐ does not meet the National Register criteria. (☐ See continuation sheet for additional comments.)

Signature of certifying official/Title Date

State or Federal agency and bureau

4. National Park Service Certification

I hereby certify that the property is:

- ☑ entered in the National Register.
  - ☐ See continuation sheet.
- ☐ determined eligible for the National Register
  - ☐ See continuation sheet.
- ☐ determined not eligible for the National Register.
- ☐ removed from the National Register.
- ☐ other, (explain:) ______

Signature of the Keeper: for Edson H. Beall

Date of Action: 10/10/02

5. Classification

Ownership of Property
(check as many boxes as apply)

- [x] public-local
- [ ] private
- [ ] public-State
- [ ] public-Federal

Category of Property
(check only one box)

- [ ] district
- [ ] building(s)
- [x] site
- [ ] structure
- [ ] object

Number of Resources within Property
(Do not include previously listed resources in the count.)

Contributing	Noncontributing	
		buildings
1		sites
		structures
		objects
1		Total

Name of related multiple property listing
(Enter "N/A" if property is not part of a multiple property listing.)

N/A

Number of contributing resources previously listed in the National Register

N/A

6. Function or Use

Historic Function
(Enter categories from instructions)

FUNERARY/Cemetery

Current Function
(Enter categories from instructions)

FUNERARY/Cemetery (inactive)

7. Description

Architectural Classification
(Enter categories from instructions)

N/A

Materials
(Enter categories from instructions)

foundation N/A
walls N/A
roof N/A
other N/A

Narrative Description
(Describe the historic and current condition of the property on one or more continuation sheets.)

Description:

The Garrison Cemetery (also known as the "War of 1812 Cemetery") is located in the modern day Town of Cheektowaga, New York. This small military cemetery is situated on the south side of Aero Drive, between Garrison Road to the west and Youngs Road on the east, with Ellicott Creek on the north. Aero Drive runs along the southern bank of Ellicott Creek, which has a west-northwest to east-southeast orientation in this area. The 0.319-acre site is bounded by a modern driveway and parking lot owned by the Niagara Frontier Transportation Authority to the east, by the Buffalo-Niagara International Airport on the south and on the west by another modern service entrance owned by the Automated Flight Service Station. Historically, this area was farmland, but developed commercially beginning with the airport construction in the 1920's.

The nominated cemetery parcel is located in the northern portion of Lot 7, Township 11, Range 7 of the Holland Land Company Purchase. The cemetery is a five sided polygon shape and is surrounded by a wood rail fence with a large wooden plaque located on the eastern perimeter facing Aero Drive. Two stone boulders with mounted bronze plaques (1933) and two cut stones from an earlier fence mark the cemetery entrance along Aero Drive. Cement markers are located outside the fence line at the corners and along the southern border of the cemetery. Three trees stand in the eastern section of the cemetery, while two stand in the western area. A commemorative stone is situated in front of a large cannon, which is located approximately 40 feet south of the entrance and is flanked by two flagpoles. The cannon, consisting of a large American fieldpiece captured by the British during the War of 1812 is mounted on a replica wooden field carriage, which rests on a raised concrete platform. The current cannon and platform that were arranged at the center of the cemetery in 1936 and replaced an earlier cannon associated with the cemetery from 1898 to 1942. There are ten small white crosses protruding from the ground in the western section, while there are twelve found east of the cannon. There are four commemorative flower beds within the cemetery. The two large bermed beds flank the cannon, a smaller bed is located to the rear of the cannon and the fourth bed is in the western section near the fence.

A geophysical survey was conducted at the Garrison Cemetery on May 31, 2001 by Geosphere Inc. of Auburn Michigan to detect and map grave sites to demonstrate that the designated cemetery was truly a mass grave site. The geophysical methods used include electromagnetics (EM) and ground penetrating radar (GPR). The following comments were incorporated into a report completed by the consultant: Analysis of the EM38 and radar results identified the location and extent of two very large disturbed areas that are interpreted as communal grave areas. One is located in the eastern portion of the Cemetery and the other encompasses the entire western part of the Cemetery. The radar results also determined the approximate depth of the excavation and edges of the burial areas. It is our (Geosphere, Inc.) interpretation that many hundreds of individuals were likely laid to rest in these two large areas. Anomalies were also located at a few of the existing marked (individual) graves with the radar.

8. Significance

Applicable National Register Criteria
(Mark "x" in one or more boxes for the criteria qualifying the property for National Register listing.)

☒ **A** Property is associated with events that have made a significant contribution to the broad patterns of our history.

☐ **B** Property is associated with the lives of persons significant in our past.

☐ **C** Property embodies the distinctive characteristics of a type, period, or method of construction or represents the work of a master, or possesses high artistic values, or represents a significant and distinguishable entity whose components lack individual distinction.

☒ **D** Property has yielded, or is likely to yield, information important in prehistory or history.

Criteria Considerations
(Mark "x" in all the boxes that apply.)

Property is:

☐ **A** owned by a religious institution or used for religious purposes.

☐ **B** removed from its original location.

☐ **C** a birthplace or grave.

☒ **D** a cemetery.

☐ **E** a reconstructed building, object, or structure.

☐ **F** a commemorative property.

☐ **G** less than 50 years of age or achieved significance within the past 50 years.

Narrative Statement of Significance
(Explain the significance of the property on one or more continuation sheets.)

Areas of Significance
(enter categories from instructions)

Military

Period of Significance
ca. 1814--ca. 1815

Significant Dates
ca. 1814--ca. 1815

Significant Persons
(Complete if Criterion B is marked above)
N/A

Cultural Affiliation
N/A

Architect/Builder
N/A

9. Major Bibliographical References

Bibliography
(Cite the books, articles, and other sources used in preparing this form on one or more continuation sheets

Previous documentation on file (NPS):

☐ preliminary determination of individual listing (36 CFR 67) has been requested
☐ previously listed in the National Register
☐ previously determined eligible by the National Register
☐ designated a National Historic Landmark
☐ recorded by Historic American Buildings Survey # ____________
☐ recorded by Historic American Engineering Record # ____________

Primary location of additional data:

☒ State Historic Preservation Office
☐ Other State agency
☐ Federal agency
☐ Local government
☐ University
☐ Other Name of repository:

Buffalo & Erie County Historical Society

Statement of Significance:

The Garrison Cemetery along Aero Drive in the Town of Cheektowaga is historically significant under Criterion A for its association with the 1814-1815 Niagara Frontier Campaign of the War of 1812 between the United States and Great Britain. The military cemetery is derives further significance for its association with the General Military Hospital established August 1, 1814 at Williams Mills, now modern day Williamsville, New York, for the garrison commanded by General Jacob Brown. Western New York was an area of important military activities during the War of 1812 and the Garrison Cemetery is one of the very few remaining historic sites associated the Niagara Campaign.

Williamsville itself was an integral part of America's war effort on the Niagara Frontier as early as 1812. General Alexander Smythe had established winter quarters along what is known as Garrison Road between Main Street (Buffalo Road) and Ellicott Creek.1 By October of 1812, the command along the Niagara Theater had grown to 6,000 men dispersed along the Niagara River from Buffalo to Fort Niagara. General Stephen Van Rensselaer commanded a force of nearly 2,000 militiamen, 2,400 regulars and 2,200 troops at Buffalo under the command of General Smythe. The Americans met with early defeats due partially to the failure of General Smythe to cooperate with General Van Rensselaer, who had wanted to attack Queenston with the bulk of the forces. General Smythe attempted to take Fort Erie in Canada and summarily failed and would find himself in a confrontation with General Peter B. Porter, a New York politician and businessman. General Smythe would eventually be relieved of duty and would no longer play a role in the war.2

Early in 1813, plans were developed to attack the British forces along the Niagara Frontier and orders were given to concentrate 3,000 regulars and volunteers at Buffalo and 4,000 at Sackets Harbor.3 The strategy was to attack York (Toronto) and after destroying the arsenal, to attack Fort George on the mouth of the Niagara River directly across from Fort Niagara. A simultaneous attack would take place at Fort Erie and Fort Chippewa with the troops stationed at Buffalo and Black Rock, the goal being to drive the British back along the line of the Niagara River from Lake Erie to Lake Ontario.4 During April and May of 1813, the American forces launched a successful attack on York and troops led by Colonel Winfield Scott and Captain Oliver H. Perry defeated the British at Fort George. Another force led by Lt. Colonel James Preston crossed the Niagara River at Black Rock and captured Fort Erie. The American Forces had effectively gained control of the Niagara River, but a series of misjudgments and failures would result in the British taking possession of Fort Erie, Fort George and Fort Niagara in December of 1813.

In October 1813, the Americans attempted an invasion of Montreal, Canada, led by General James Wilkinson. A battle ensued on the St. Lawrence River, with the temperature below freezing and snow and sleet falling, resulting in the American Forces retreating until reinforcements could join them. The American Army was to move no further, establishing winter quarters in French Mills. The assault led by General James Wilkinson was a disaster, not only because he failed, but also would prove calamitous to the American Forces on the Niagara Frontier. General John Vincent, who had been stationed on the Niagara departed with 800 regulars to join General Wilkinson, leaving General McClure at Fort George with a greatly reduced force of poorly trained troops. As General McClure would report to the Secretary of War, there were only "sixty effective regular troops and probably forty volunteers."5 On December 10, 1813, General McClure deserted Fort George, burning it in his wake. He would also set fire to the Town of Newark (Niagara-on-the-Lake), destroying the entire village and leaving the inhabitants without shelter for the winter.

The British were quick to retaliate under the command of Lt. General Gordon Drummond who approved a plan that would send two forces across the Niagara River to attack the American Forces under the command of Colonel Murray and Brigadier General Phinias Riall. The British would take possession of Fort George and Fort Niagara and would burn all the settlements between Fort Niagara and Tonawanda Creek. The "New York Evening Post" would report that the "enemy crossed the Niagara River at Lewiston with 2,000 men, which place they burned and marched up to Schlosser over the falls, burning and destroying every house on their route"6. On December 28th and 29th the British attacked Black Rock and Buffalo and these villages would subsequently be burnt. The American forces were in complete disarray, as well as the residents of Buffalo. Williamsville, the village eleven miles from Buffalo would become their sanctuary. The "Manilus Times" would report, "the inhabitants are scattered in every direction. Most of whom have come to Williamsville...where our force is assembled"7 The local citizens and American soldiers fled Buffalo along the Buffalo Road to Williamsville, the only settlement of any size between Buffalo and Batavia in Genesee County. This small community of twenty or more buildings has already become a military garrison in the fall of 1812 due to the dangerous situation on the Niagara Frontier and the approaching winter. The soldiers kept busy building log barracks along the Buffalo Road and south toward Ellicott Creek along what is now known as Garrison Road. The sick and wounded from Lewiston were moved into these improved barrack hospitals.

Williamsville would become the center of operations for the military following the destruction of Buffalo, as a company of cavalry under the command of Brigadier General Hopkins was dispatched from Batavia in January of 1814.8 General orders were issued to Brigadier General Burnett, giving him command of the troops and ordering him to repair the cantonment at Williamsville and construct additional housing to take care of the troops. 9 On January 31, General Amos Hall's headquarters were established at Williamsville and he would proceed to send a letter of gratitude to Brigadier General Burnett, Lt. Colonels Davis and Colt and other officers "for their exertions in regulating the cantonment, building and repairing huts, and their indefatigable exertions to render the situation of the soldiers as comfortable as possible" 10 The spring of 1814 saw a major build-up of the American forces as they prepared for an invasion of Canada. General Jacob Brown was appointed the new commander of American Forces replacing Wilkinson, with the focus on conquering the Niagara Frontier. General Brown would have an aggregate of 3,500 men, which included the First Brigade commanded by General Scott with 1,377 men, the Second led by General Eleazer Ripley included 1,128 men, the New York State Militia under the leadership of General Peter Porter numbered 600, excluding the Native Americans and the artillery numbered 345. 11

On July 3, 1814, General Jacob Brown ordered the attack on Fort Erie, Canada and General Scott advanced immediately, followed later by General Ripley. The British capitulated with very little conflict, resulting in four Americans and one British solder dead in the skirmish. 12 Brigadier General Riall, after learning of the attack on Fort Erie, assembled the British reinforcements and marched them toward Chippewa.

On July 5, 1814, General Porter and the militia marched to join General Scott at Fort Erie. General Porter encountered the British, whom he attacked, not realizing that he was only a few miles from the core of the British forces. When he recognized this, he retreated to join forces with the main body under General Scott. General Riall led his British command towards the Chippewa Bridge where he encountered General Scott's forces and a major battle ensued. Both sides claimed victory, although both armies remained in the same positions held before the battle. The casualty for both sides was great. The American casualties totaled 60 killed, 249 wounded, and 19 missing; the British Forces lost 148, with 211 wounded and 46 missing. 13 On July 24, 1814, General Brown and the remainder of his American army arrived at Chippewa and on the following day he ordered General Scott to cross the Chippewa Bridge, under the assumption that the British forces had retreated. The British had not but instead, they had established their forces at Lundy's Lane. By that evening General Scott's small brigade had encountered the entire force under the command of British General Drummond. General Scott managed to hold them off until General Ripley's command and General Porter's militia finally joined his forces. Again both sides claimed victory and again the casualties were enormous. The

American casualties totaled 171 killed, 570 wounded, and 17 missing. There were 84 British killed, 554 wounded and 235 missing. 14 General Brown and General Scott were both wounded during this battle. Major General Phinias Riall, the British Commander, was also wounded and captured by the American forces.

On August 3, 1814, the British dispatched six hundred men under the command of Lt. Colonel John Tucker across to the American side of the Niagara River. His intention was to destroy the stores of ammunition and supply depots at Buffalo and Black Rock. Major Morgan, under General Brown's command, repelled the British troops and the assault failed. The British continued their assault on Fort Erie. They bombarded the fort between August 7 and August 14, 1814. On August 15, 1814, an all out assault by the British occurred. Two of the three British columns, which marched upon the fort, were unsuccessful. Colonel Drummond was more successful, when a destructive incident occurred. The platform upon which his men were attacking from covered a storage area for a large quantity of ammunition. Suddenly a tremendous explosion took place, which blew up the magazine and the men upon it. Panic seized the British forces and they retreated, thus ending this conflict. The British casualties totaled over 900, many of who were wounded and taken prisoner. 15

The American Forces secured Fort Erie and General Orders were issued providing for the requisition of one thousand men for General Brown. Additional men were also summoned from General Hall's command. These forces were directed to rendezvous at Williamsville by September 1, 1814. 16 The reinforcements were to be sent to Fort Erie and on September 17, 1814, another bloody battle ensued, with British casualties numbering approximately eight hundred and the American losses totaling over five hundred. 17 The Americans held Fort Erie until November 5, 1814, when they set off an explosion and destroyed it. The American forces then withdrew from Fort Erie to the Village of Buffalo and Williamsville, where they established winter quarters. On December 24, 1814, the Treaty of Ghent was signed ending the War of 1812, although one more battle was to be fought. American forces commanded by General Andrew Jackson defeated the British at New Orleans on January 8, 1815. The treaty provided for the mutual restoration of territory captured by both sides.

During this period of deadly battles in 1814, it had become apparent that there was an increasing need for medical facilities to care for the sick and wounded. Prior to the July invasion of Canada, General Scott had periodically inspected the various regimental field hospitals, advising the surgeons and medical staff that the upcoming campaign would result in an increase of work for them. 18 At the end of July 1814, after fighting the Battles of Chippewa and Lundy's Lane, the medical facilities received the wounded as well as the sick from these engagements. Because of a

British raid on August 3, 1814 near the Scajaquada Creek, General Scott decided to concentrate the medical facilities at Williamsville.

Mr. Joseph Whitehorne, in his extensive research of the War of 1812, states that Senior Surgeon, Dr. Ezekiah Bull, selected the site for a general medical hospital on July 29, 1814 and Dr. Joseph Lowell was assigned to command the Williamsville General Hospital facility. "Thus, 90 acres and the stables of Raphael Cook's farm were leased for the construction of a general hospital. Ironically, the contract specified "no burying place in the premises." 19 The existing cemetery along Aero Drive was located in the vicinity of the hospital, which was consistent with burying practices of the period. The exact location of the hospital is not known, because Raphael Cook had not yet paid off his mortgage to the Holland Land Company and it is presumed that the assessment records for this period were destroyed during the burning of Buffalo.

Besides the military engagements in July of 1814, the weather and health conditions would also play an important role in the deaths of many of the American soldiers. The Buffalo Hospital served as a clearing center for the soldiers from Fort Erie, sending the men to Williamsville as soon as possible. The number of sick ranged between three and four hundred and the number of wounded much greater. 20 The rainy season beginning in August 1814 and extending into the fall, where it continued to be uncommonly warn, would result in diarrhea, dysentery and typhus, the latter two, which were extremely severe. Dr. Joseph Lowell, in his report, indicated that patients were not sent to the general hospital until at least the third week of the disease, after being treated by various methods, depending on the surgeon in charge. These diseases are said to have generally accounted for more deaths than trauma or wounds, and drug overdoses could have also accounted for a number of deaths. 21 The War Department Records discovered in 1932 exemplified this premise as these men had died from diarrhea, pneumonia, dysentery, typhus, rheumatism and other such diseases. 22

The small military cemetery in Lot 7, Township 11, Range 7 was originally purchased as part of 202 acres of land from the Holland Land Company on October 8, 1808 by Frank Ludlow, who owned it until January 11, 1816. Based on War Department Records, this cemetery along Aero Drive became a mass grave for American Soldiers beginning with William Hewes on August 4, 1814 and ending with William Pettingall July 16, 1815. It is most likely that many more American soldiers were buried in the mass grave, but the records do not survive. An unknown number of British soldiers captured during the Niagara Campaign of 1814 were also buried here. During the period in which soldiers were being buried, Ensign Jacob Porter Norton on duty with

---

the 4th Infantry Regiment wrote in his diary on May 31, 1815 that, "The army burying ground at this place when we first came, was in a miserable situation,--Many of the graves were not more than half filled, and they had been dug promiscuously without regard to order, so that it was difficult in some places to distinguish the graves. Lt. Colonel Ranney had them mounded up and clodded with green sard, the stumps dug up, a fence built round it, a gate with an arch over the way, upon the arch was inscribed--"Sacred to the memory of those men who died of wounds received in the memorable fields of Chippewa, Bridgewater and Erie,--They rest in honour and deserve the gratitude of their country". 23

In 1932, the Buffalo Historical Society received a copy of two volumes of a hospital register that were discovered in the War Department Records at the National Archives in Washington, D. C. The soldiers buried in this cemetery along Aero Drive were members of the 1st, 4th, 5th, 9th, 10th, 11th, 14th, 15th, 16th, 17th, 19th, 21st, 22nd, 23rd and 25th Regiments of Infantry, Corps of Artillery, Light Dragoons, New York Volunteer Cavalry and the New York and Pennsylvania Volunteers. The soldiers' names are available at the Buffalo and Erie County Historical Society and also include names of soldiers who they could not determine whether or not they were buried at this military cemetery along Aero Drive.

The names of the British soldiers buried in the cemetery have not been found or do not exist. An early account of the transfer of wounded British soldiers following the battle at Fort Erie on August 14, 1814 has been documented. William Hodge had been an eyewitness to the transport of British wounded to the hospital at Williamsville. "I remember well…they were in great pain, and begged for whiskey or some other liquor to drown their suffering. They were awful objects to behold--some with their eyesight gone, and others with their faces blackened and other wist disfigured". 24 Benson J. Lossing in his book Pictorial Field-Book of the War of 1812, also describes the transferring of British prisoners to Williamsville. 25

In January 1876, Timothy A. Hopkins, son of Brigadier General Timothy S. Hopkins, New York State Militia stationed at Williamsville, received a letter from Alex J. Sheldon of Buffalo inquiring about the War of 1812 cemetery located in Williamsville, because the War Department had asked for a report of burials. Hopkins response states: "The Hospital Barracks were situated about one mile above the Village, on the bank of Ellicott Creek, and a short distance above is their place of burial, covering less than one acre of ground, upon which two hundred or more were buried, some in separate graves, others several in one grave, without coffins, and no possibility of identifying any, where they have lain for sixty years forgotten by their friends, the busy world and the government they sacrificed their lives to sustain, and upon the south east corner was also buried in one common grave the English prisoners fatally injured at the affair of

Fort Erie. The grounds are not fenced except on the street leading up the Creek, the lands adjoining have been continually under cultivation and some think the occupants have encroached upon the grounds of this I cannot say." 26 Three years later, in 1879, William Hodge and Timothy A. Hopkins visited the cemetery together. Mr. Hodge's description of the cemetery in 1897 indicates that the area that comprised the cemetery has not changed much since that time. He describes it as being about a half acre, on the southwesterly side of a public road, parallel to what was then called Eleven-Mile Creek. He further states that the "ground is a little elevated from the surrounding land and road adjoining. On the border of the two sides stand ten sugar maple trees...I have no doubt they were planted by the comrades of those whose bones hear occupy this land." 27 The chronological documentation describes, locates and places American and British soldiers in the cemetery.

During the period of burial between 1814 and 1815, Frank Ludlow owned the property where the Garrison Cemetery is located. The land passed through several owners until 1851 when Captain John Bliss and John R. Evans purchased the cemetery from the Haskel family to keep it intact and to save it from complete deterioration. It was passed down to members of their families until 1898. The cemetery was then transferred to the Buffalo Historical Society and the deed stated that "said premises to remain and be perpetually a cemetery and resting place for the dead soldiers of the War of 1812" 28 An 1898 newspaper article and photograph of the cemetery appeared in the Buffalo Express. The cemetery was described as "long neglected" and lined with a "row of fine large maples on one side...and Ellicott Creek runs in front".29 The attached photograph (1898) shows the newly placed cannon and the metal fence with an arched gateway having letters cut from galvanized iron, which read "Soldiers of the War of 1812" The ca. 1900 photographs clearly shows the arch, which is barely visible in the 1898 photo. In 1919 a newspaper article appeared in the Amherst Bee that describes the cemetery as an area that time has caused "havoc" and that "the fence has partially fallen and dilapidation and ruin have set their mark there". 30 The attached photo dated 1919 appeared with the article and shows the described condition.

The attached historic photographs reinforce the existence of the cemetery, which originally had no markers. The recent electromagnetics (EM) and ground penetrating radar (GPR) clearly identify the existence and location of two very large communal grave areas. The attached site plan shows a slight depression on the west, which is the large mass grave for the American Soldiers, while the depression on the east is the large mass grave for the British Soldiers.

In 1985 the Town of Cheektowaga acquired the Garrison Cemetery along Aero Drive from the Buffalo and Erie County Historical Society. The Cheektowaga Town Board continues to maintain the cemetery and to hold annual ceremonies to pay tribute to all those soldiers who served their respective countries.

The Garrison Cemetery is significant for its association with the War of 1812 and is an important local landmark recalling a difficult period in American history.

Footnotes:

1 Bingham, Robert W. (1947). A Brief History of the Village of Williamsville. Amherst Bee. Amherst, New York.

1 Whitehorne, Joseph (1992). While Washington Burned, The Battle for Fort Erie 1814. Balltimore, Maryland, The Nautical & Aviation Publishing Company of America, p. 40.

3 Bierne, Francis F. (1949). The War of 1812. New York, E. P. Dutton & Co., p. 157-158.

4 Ibid., p. 158.

5 Cruikshank, Ernest. Editor (1971). The Documentary Histori of the Campaign of the Niagara Frontier, 1812-1814. New York, Arno Press & New York Times. "American State Papers, Military Affairs", Vol. 1, p. 486.

6 Ibid., "New York Evening Post", January 11, 1814., Vol. III, Part 2, p. 83.

7 Ibid., "Manilus Times", January 4, 1814, Vol. III, p. 83.

8 Severence, Frank H. (1902). Papers Relating to the War of 1812 on the Niagara Frontier. Publication of the Buffalo Historical Society. Buffalo, Buffalo Historical Society. V, pp. 41-42.

9 Ibid., pp. 47-48.

10 Ibid., p. 49.

11 Bierne, p. 251.

12 Ibid., p. 253.

13 Babcock, Louis L. (1899). The Siege of Fort Erie, An Episode of the War of 1812. Buffalo, The Peter Paul Book Co., p. 23.

14 Ibid., p. 23.

15 Ibid., p. 63.

16 Cruikshank, "State of New York, General Order", "Buffalo Gazette", Vol. IV, Part 2, pp. 438-439.

17 Babcock, p. 63.

18 Whitehorne, p. 50.

19 Whitehorne, p. 55.

20 Mann, Dr. James. (1815). Medical Sketches of the Campaign of 1812, 1813, 1814. H. Mann & Co., pp. 94-96.

21 Noe, Adrianne (1993) Medical Records. Snake Hill, An Investigation of a Military Cemetery from the War of 1812. Susan Pfieffer, Ronald F. Wilkinson, Editors, Toronto and Oxford, p. 67

22 National Archives Washington, D. C. (1814-1815) Register of Patients in the Hospital at Williamsville, Entry 56-9W3/17/11/1 Volume 2 and 3. Washington, D. C.

23 Porter, Daniel R., ed. (1814) Jacob Porter Norton, A Yankee on the Niagara Frontier in 1814. Niagara Frontier. 12: pp.51-57.

24 Hodge, William (1922). Recalling Pioneer Days. The Buffalo Historical Society Publications. F. H. Severance. Buffalo, New York, The Buffalo Historical Society. XXVI, p. 229.

25 Lossing, Benson J. (1869). Pictorial Field-Book of the War of 1812. New York, Harper & Brothers, Publishers, (http://freepages.history.rootsweb.com~wearr1/Lossing2/Chapter35.html).

26 Hoopkins, Timothy A. (February 10, 1876). Letter Regarding War of 1812 Cemetery at Williamsville. Alex Sheldon. Williamsville, New York, Buffalo and Erie County Historical Society.

27 Hodge, William (1879). Buffalo Cemeteries. Publication of the Buffalo Historical Society. Buffalo, New York, The Buffalo Historical Society. I: pp. 68-70.

28 Monroe Abstract & Title Corporation (1985). Complete Record Search #613920, Search of property known as the War of 1812 Cemetery.

29 Soldiers' Burying-Ground at Williamsville Saved From Neglect". Buffalo Express. Buffalo, New York, (1898).

30 "Soldiers' Plot to be Improved. " Amherst Bee. (1918).

10. Geographical Data

Acreage of Property **0.319 Acre**

UTM References
(Place additional boundaries of the property on a continuation sheet.)

1 17
Zone Easting Northing

2
Zone Easting Northing

3
Zone Easting Northing

4
Zone Easting Northing

☐ See continuation sheet

Verbal Boundary Description
(Describe the boundaries of the property on a continuation sheet.)

Boundary Justification
(Explain why the boundaries were selected on a continuation sheet.)

11. Form Prepared By

name/title Claire L. Ross, Program Analyst

organization NYS Office of Parks, Recreation and Historic Preservation date June 10, 2002

street & number Peebles Island, P.O. Box 189 telephone 518-237-8643

city or town Waterford state NY zip code 12188-0189

Additional Documentation

Submit the following items with the completed form:

Continuation Sheets

Maps

A **USGS map** (7.5 or 15 minute series) indicating the property's location.
A **Sketch map** for historic districts and properties having large acreage or numerous resources.

Photographs

Representative black and **white photographs** of the property.

Additional items
(Check with the SHPO or FPO for any additional items)

Property Owner

(Complete this item at the request of SHPO or FPO.)

name/title Town of Cheektowaga

street & number 3301 Broadway telephone 716-686-3433

city or town Cheektowaga state N Y zip code 14227

Paperwork Reduction Act Statement: This information is being collected for applications to the National Register of Historic Places to nominate properties for listing or determine eligibility for listing, to list properties, and to amend existing listings. Response to this request is required to obtain a benefit in accordance with the National Historic Preservation Act, as amended (16 U.S.C. 470 *et seq.).*

Estimated Burden Statement: Public reporting burden for this form is estimated to average 18.1 hours per response including time for reviewing instructions, gathering and maintaining data, and completing and reviewing the form. Direct comments regarding this burden estimate or any aspect of this form to the Chief, Administrative Services Division, National Park Service, P.O. Box 37127, Washington, DC 20013-7127; and the Office of Management and Budget, Paperwork Reductions Projects (1024-0018), Washington, DC 20503.

Verbal Boundary Description:

The boundary for the nominated property is shown as the heavy black outline on the accompanying Survey for the War of 1812 Cemetery, Town of Cheektowaga, Erie County, New York.

Boundary Justification:

The boundary for the nomination has been drawn to coincide with the current parcel for the property

Consultant:

Original Research and First Draft Nomination:

Ms. Mary F. Holtz
Town of Cheektowaga Historian
3301 Broadway
Cheektowaga, New York 14227

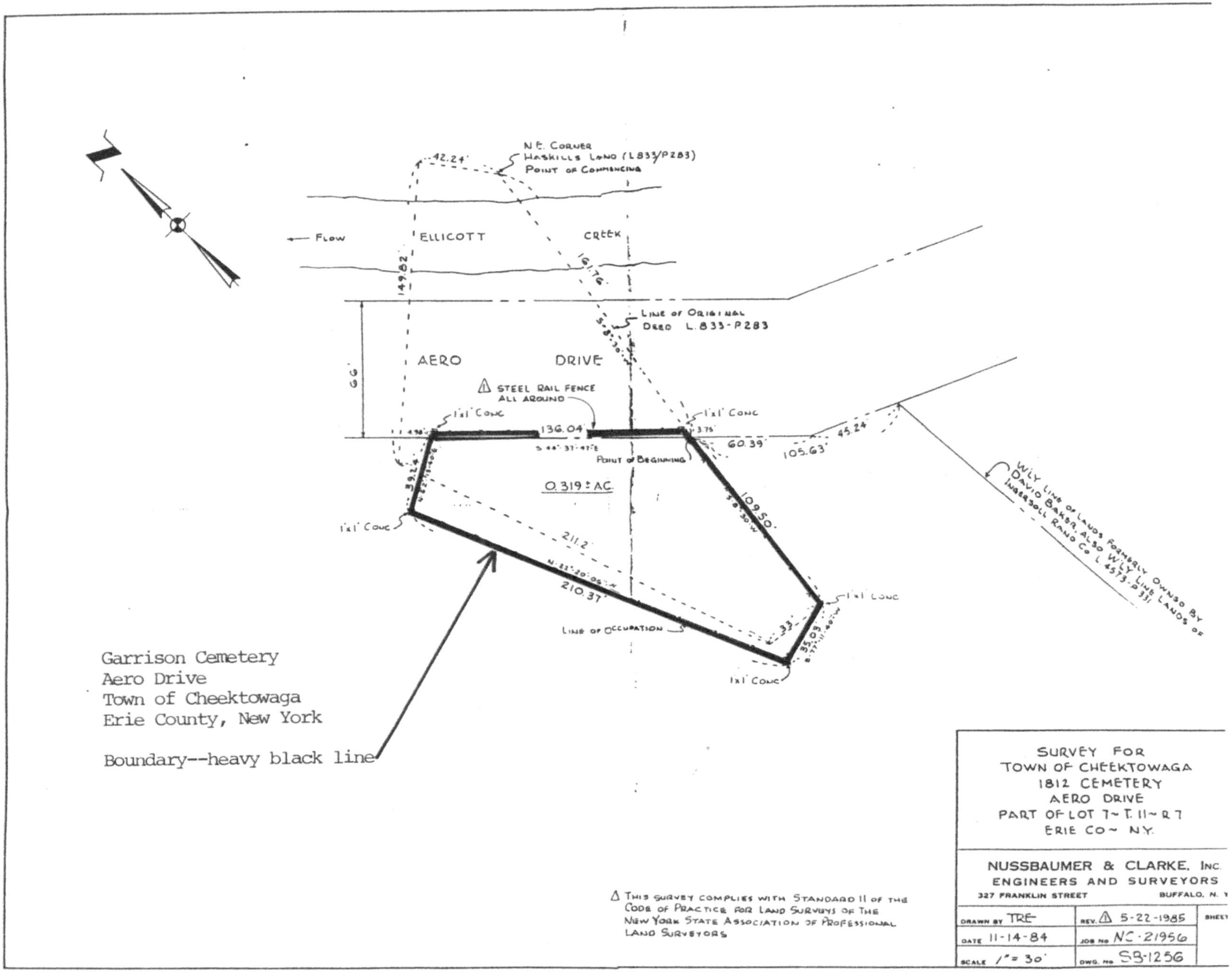
N.E. Corner
Haskills Land (L833/P283)
Point of Commencing
42.24'
Flow
Ellicott
Creek
149.82'
16.76'
Line of Original
Deed L.833-P283
66'
Aero
Drive
Steel Rail Fence
All Around
1x1' Conc
136.04'
Point of Beginning
60.39'
105.63'
45.24'
0.319± AC.
109.50'
211.2'
210.37'
Line of Occupation
33'
35.09'
Wly Line of Lands Formerly Owned By
David Baker, Also Wly Line Lands of
Ingersoll Rand Co L 4573-P331
Garrison Cemetery
Aero Drive
Town of Cheektowaga
Erie County, New York
Boundary--heavy black line
Survey for
Town of Cheektowaga
1812 Cemetery
Aero Drive
Part of Lot 7~T. 11~R 7
Erie Co~ N.Y.
Nussbaumer & Clarke, Inc.
Engineers and Surveyors
327 Franklin Street
Buffalo, N. Y.
Drawn by TRE
Rev. 5-22-1985
Date 11-14-84
Job No NC-21956
Scale 1" = 30'
Dwg. No SB-1256
This survey complies with Standard II of the
Code of Practice for Land Surveys of the
New York State Association of Professional
Land Surveyors

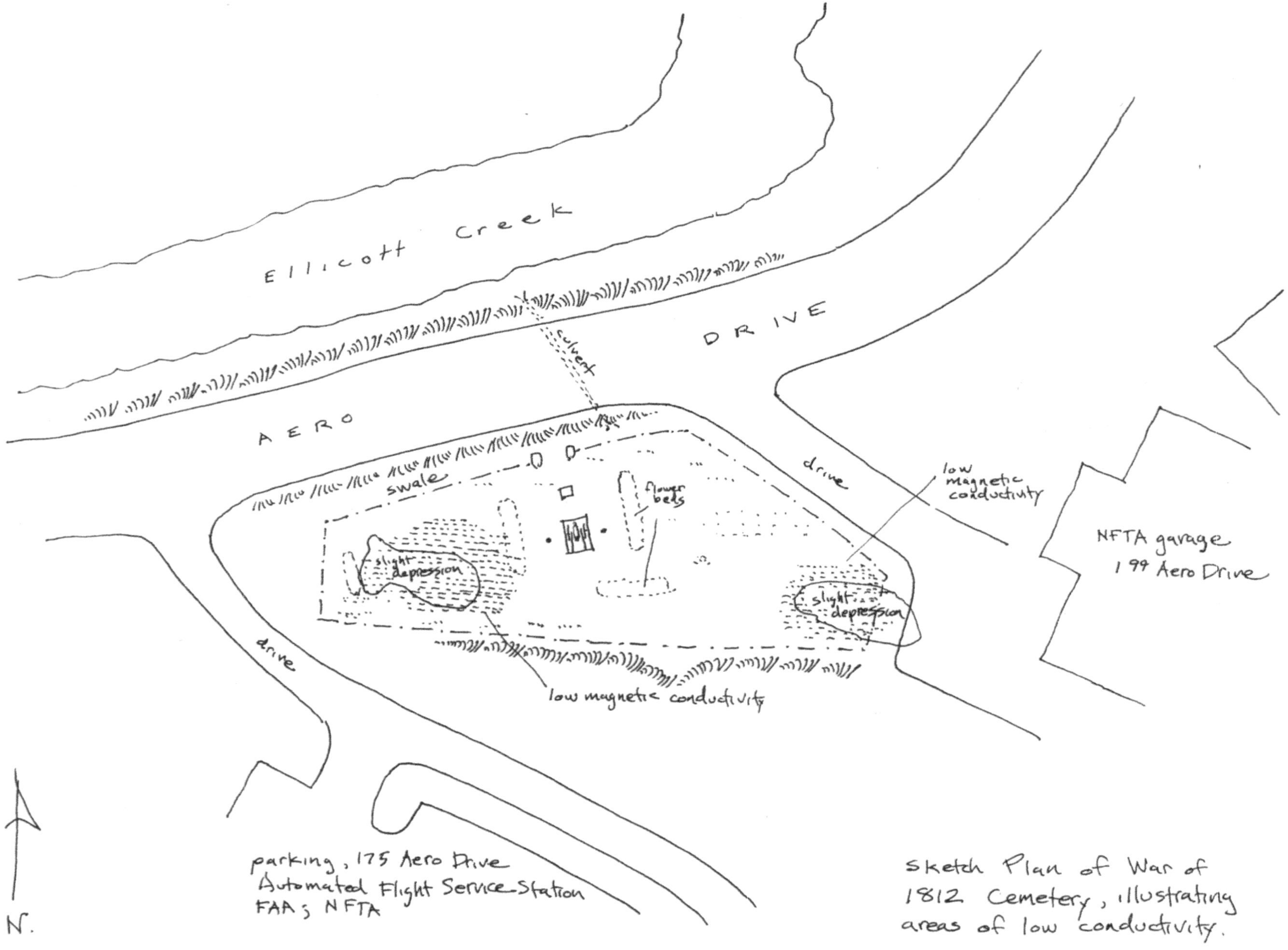

Sketch Plan of War of 1812 Cemetery, illustrating areas of low conductivity.

Garrison Cemetery
Aero Drive
Cheektowaga, N.Y.
Erie County

Scale: 1:24,000
Lancaster Quad

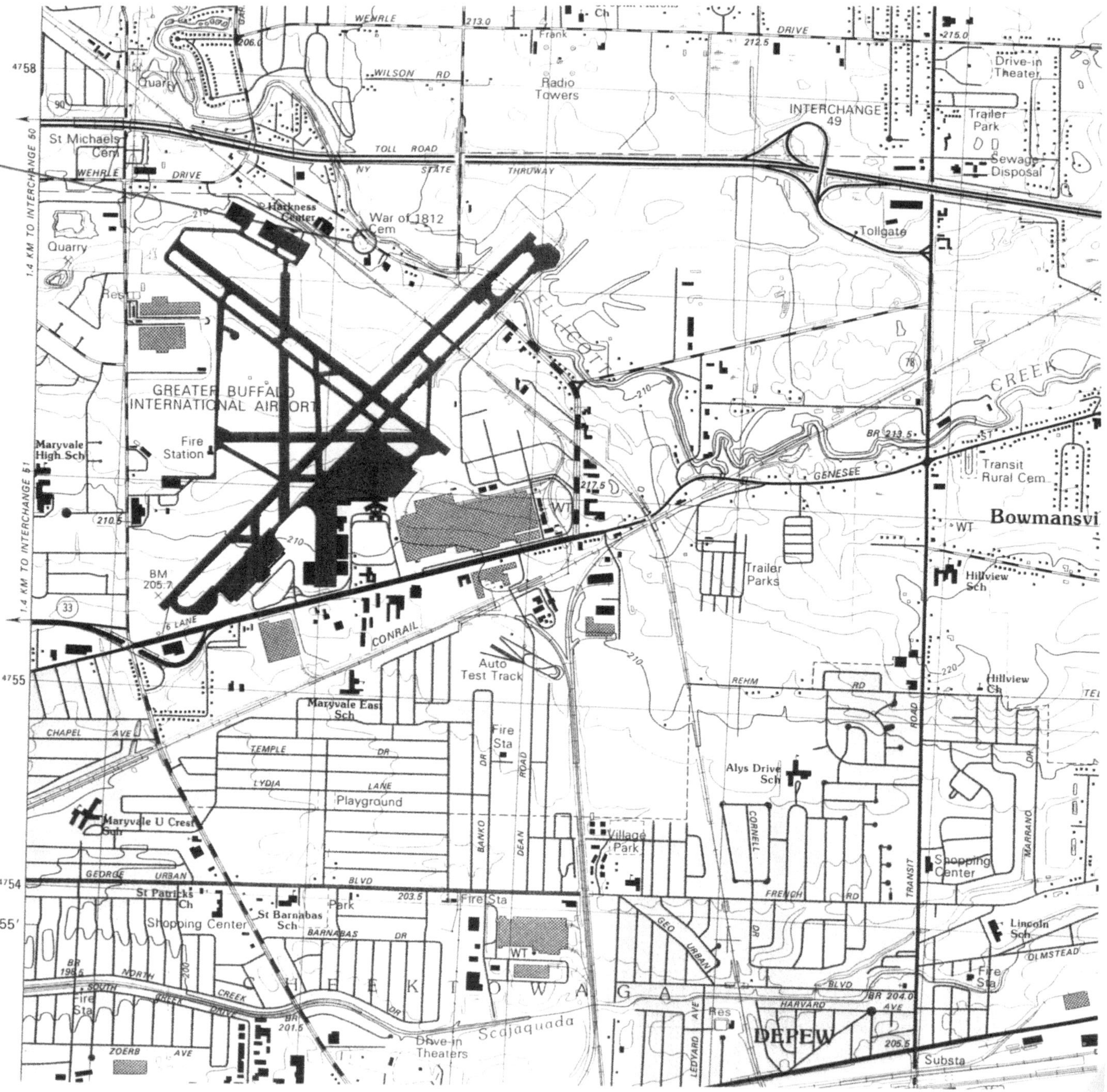

WAR OF 1812 CEMETERY
AERO DRIVE, CHEEKTOWAGA, NEW YORK

Appendix N - Photograph of Cemetery - 1898

Appendix N - Photograph of Cemetery - Circa 1900

WAR OF 1812 CEMETERY
AERO DRIVE, CHEEKTOWAGA, NEW YORK

OLD GARRISON BURYING GROUND.

APPENDIX N - PHOTOGRAPH OF CEMETERY - 1919

WAR OF 1812 CEMETERY
AERO DRIVE, CHEEKTOWAGA, NEW YORK

APPENDIX O - PHOTOGRAPH OF CEMETERY - 1939

UNITED STATES DEPARTMENT OF THE INTERIOR
NATIONAL PARK SERVICE

NATIONAL REGISTER OF HISTORIC PLACES
EVALUATION/RETURN SHEET

REQUESTED ACTION: NOMINATION

PROPERTY NAME: Garrison Cemetery

MULTIPLE NAME:

STATE & COUNTY: NEW YORK, Erie

DATE RECEIVED: 8/27/02 DATE OF PENDING LIST: 9/23/02
DATE OF 16TH DAY: 10/09/02 DATE OF 45TH DAY: 10/11/02
DATE OF WEEKLY LIST:

REFERENCE NUMBER: 02001113

REASONS FOR REVIEW:

APPEAL:	N	DATA PROBLEM:	N	LANDSCAPE:	N	LESS THAN 50 YEARS:	N
OTHER:	N	PDIL:	N	PERIOD:	N	PROGRAM UNAPPROVED:	N
REQUEST:	N	SAMPLE:	N	SLR DRAFT:	N	NATIONAL:	N

COMMENT WAIVER: N

_✓_ACCEPT ___RETURN ___REJECT 10/10/02 DATE

ABSTRACT/SUMMARY COMMENTS:

Entered in the National Register

RECOM./CRITERIA__________________

REVIEWER__________________ DISCIPLINE__________________

TELEPHONE__________________ DATE__________________

DOCUMENTATION see attached comments Y/N see attached SLR Y/N

THE ASSEMBLY
STATE OF NEW YORK
ALBANY

PAUL A. TOKASZ
Assemblyman 143rd District

DISTRICT OFFICE:
General Donovan State Office Building
125 Main Street
Buffalo, New York 14203
(716) 852-2791
FAX (716) 852-2794

ALBANY OFFICE:
Room 727
Legislative Office Building
Albany, New York 12248
(518) 455-5921
FAX (518) 455-3962

CHAIRMAN
Committee on Election Law

COMMITTEES
Environmental Conservation
Higher Education
Local Governments
Transportation

July 7, 1995

Commissioner Bernadette Castro
NYS Department of Parks,
Recreation and Historical Preservation
Empire State Plaza
Agency Building 1
Albany, New York 12238

Dear Commissioner Castro:

I am writing to express my support for the Town of Cheektowaga's application to place the War of 1812 Cemetery listed on the State and Federal Registers of Historic Places.

It is my understanding that the War of 1812 Cemetery meets the requirements to be placed on the register. It is a historical landmark not only for the town, but for the state and country as well. Those who are buried in the cemetery are not only from Cheektowaga, but represent all areas of the Eastern United States.

The site also contains a War of 1812 cannon, and other memorabilia which are a mirror into our military past. By having the War of 1812 Cemetery listed on the State and Federal Registers of Historic Places, it will continue to raise the awareness of citizens everywhere that United States' ground is not immune to war.

Again I would like to offer my strong support of the Town of Cheektowaga's application for the War of 1812 Cemetery. Thank you in advance for your consideration.

Sincerely,

Paul A. Tokasz
Member of Assembly

PAT/rj

cc: Honorable Thomas Johnson Jr.
Deputy Town Attorney Kevin G. Schenk
Nancy Todd

JACK QUINN
30TH DISTRICT, NEW YORK

TRANSPORTATION AND
INFRASTRUCTURE

SUBCOMMITTEES:
SURFACE TRANSPORTATION
WATER RESOURCES AND ENVIRONMENT
RAILROADS

VETERANS' AFFAIRS

SUBCOMMITTEE:
HOSPITALS AND HEALTHCARE

JOINT ECONOMIC COMMITTEE

Congress of the United States
House of Representatives
Washington, DC 20515–3230

PLEASE RESPOND TO:

WASHINGTON OFFICE:
331 CANNON BUILDING
WASHINGTON, DC 20515
(202) 225-3306
FAX: 226-0347

MAIN OFFICE:
403 MAIN STREET
SUITE 240
BUFFALO, NY 14203-2199
(716) 845-5257
FAX: 847-0323

SATELLITE OFFICE:
1490 JEFFERSON AVENUE
BUFFALO, NY 14208
(716) 886-4076

July 13, 1995

Ms. Nancy Todd
Field Service Representative
NYS Office of Parks, Recreation and Historic Preservation
Peebels Island
P.O. Box 189
Waterford, New York 12188-0189

Dear Ms. Todd:

I am writing in support of the efforts taken by the Town of Cheektowaga to have the War of 1812 Cemetery located within the town listed on the state and federal registers of historic places.

I fully endorse the State's nomination to the national register and offer any assistance that may be necessary to expedite this process with the Secretary of the Interior. Please keep me informed of the status of the nominating process.

Thank you for your consideration of this matter. Please feel free to contact me if I can be of any assistance.

Very truly yours,

Jack Quinn
Member of Congress

JQ:ck

4-23-02
cShull
[handwritten initials] 4/25/02

APR 29 2002

H32(2280)

Honorable Jack Quinn
House of Representatives
Washington, D.C. 20515

Dear Mr. Quinn:

Thank you for your letter of March 28, 2002, to Secretary Norton seeking Federal historic recognition for the War of 1812 Cemetery in Cheektowaga, New York.

We have reviewed the materials you provided and consulted with the New York State Office of Parks, Recreation and Historic Preservation regarding the property's historic status. The cemetery may qualify for listing in the National Register of Historic Places, the official list of the Nation's cultural resources worthy of preservation. Nominations to the National Register are made by the State through the State Historic Preservation Officer, Ms. Bernadette Castro. Her address and telephone number are:

Ms. Bernadette Castro, SHPO
Parks, Recreation & Historic Preservation
Agency Building #1, Empire State Plaza
Albany, New York 12238
518-474-0443

We urge the Town of Cheektowaga to work with the staff of New York State Office of Parks, Recreation and Historic Preservation to complete a National Register nomination. National Register documentation provides the detailed information on the history and integrity of historic places so that their significance may be evaluated. Enclosed is a brochure which explains the National Register program.

We appreciate your interest in the preservation of America's historic places. If you should have further questions about the National Register of Historic Places, please contact Ms. Carol Shull, Keeper of the National Register of Historic Places, at 202/343-9504.

Sincerely,

Katherine H. Stevenson

Katherine H. Stevenson
Associate Director, Cultural Resource
Stewardship and Partnerships

Enclosure

cc: New York/SHPO
bcc: 0001
2200
2250
NHL/NERO
2605 D. Lowery (02-00434)
Basic File Retained In 2280
FNP:JSprinkle:OP:04/23/02:f/NR-NHL/Sprinkle/1812Cemetely1.wpd

FACSIMILE

United States Department of the Interior
National Park Service
Office of the Associate Director,
Cultural Resource Stewardship and Partnerships
1849 C Streets, NW, Room 3128
Washington, DC 20240-0001
(202) 208-7625
(202) 273-3237 (FAX)

Date: 4/11/02

To: Carol Shull

Fax Number:

From: Bobbie

Subject: Controlled Correspondence

Pages Transmitted 4 (including cover sheet)

Comments:

Deputy Associate Director	(202) 208-4278
National Center for Cultural Resource Stewardship	
& Partnership Programs	(202) 343-3388
Heritage Preservation Services	(202) 343-9574
Park Historic Structures & Cultural Landscapes	(202) 343-8153
Museum Management	(202) 343-8142
National Register, History, & Education	(202) 343-9536
HABS/HAER	(202) 343-9618
Archaeology & Ethnography	(202) 343-4101
National Center for Recreation & Conservation	(202) 565-1180
National Designations	(202) 565-1175
Trails and Conservation Assistance	(202) 565-1190
Recreation Program	(202) 565-1140
National Center for Preservation Technology & Training	(318) 356-7444
Federal Preservation Institute	(202) 343-9569
Partnership Office	(202) 208-5477
Heritage Areas	(202) 565-1179

WASO Administrative Program Center 202/501-1111

WAPC#:	**2002-00434**	Org Name:	**House of Repres.**
ESTP:	**313991**	Constituent Name:	**Kevin G. Schenk**
Doc Name:	**Jack Quinn**	Logged By:	**M.McBryde**
Sig Level:	**A/D**	Analyst:	**Doris Lowery**
Legislative Referral:	**ES Mail**	Type Corr:	**Secy Congressional**
Writer Desc:	**U.S. House/Single**	Issue Ref:	**Historic Landmarks**
		Opinion:	**Concerned**

Park/Office	**National Park Service**	Corr Date:	**03/28/2002**
Office Assigned:	**Cultural Resources**	WAPC Date:	**04/10/2002**
Contact Name:	**Margaret Triebsch**	Draft Due Date:	
Contact Phone:	**202/208-7625**	Final Due Date:	**04/24/2002**
Contact Fax:		Response Date:	4-22

Hot Topics:

Description: **Requests assistance in getting the War of 1812 Cemetery designated as a Historic Landmark.**

[signature]

PRIORITY

PRIORITY

Recd 4/10/02

J...K QUINN
30TH DISTRICT, NEW YORK

COMMITTEES:

TRANSPORTATION AND INFRASTRUCTURE
CHAIRMAN, SUBCOMMITTEE ON RAILROADS
SUBCOMMITTEE ON AVIATION
SUBCOMMITTEE ON HIGHWAYS AND TRANSIT

VETERANS' AFFAIRS'
SUBCOMMITTEE ON BENEFITS

CHAIRMAN, NORTHEAST MIDWEST COALITION
CHAIRMAN, STEEL CAUCUS EXECUTIVE COMMITTEE

Congress of the United States

House of Representatives

Washington, D.C. 20515-3230

PLEASE RESPOND TO:

WASHINGTON OFFICE:
☐ 2448 RAYBURN BUILDING
WASHINGTON, DC 20515
(202) 225-3306
FAX: (202) 226-0347

MAIN OFFICE:
☐ 403 MAIN STREET
SUITE 240
BUFFALO, NY 14203-2199
(716) 845-5257
FAX: (716) 847-0323

SATELLITE OFFICE:
☐ 1490 JEFFERSON AVENUE
BUFFALO, NY 14208
(716) 886-4076

March 28, 2002

The Honorable Gale A. Norton
Secretary
Department of the Interior
1849 C St Room 6156
Washington, D.C. 20240

Dear Secretary Norton,

I have recently been contacted by the Town of Cheektowaga regarding the designation of property known as the War of 1812 Cemetery as a Historic Landmark on the NYS and Federal Registries.

I have taken the liberty of enclosing the correspondence I received relative to this matter, and look forward to your expedited review and response.

This truly historic site deserves your every consideration. Your assistance in obtaining the designation will be greatly and personally appreciated.

Thank you in advance for your attention to this matter. If you have any questions or suggestions, please do not hesitate to contact me in my Buffalo district office.

Very truly yours,

Jack Quinn
Member of Congress

JQ:mc

RECEIVED
02 APR -8 PM 3:00
OFFICE OF THE
EXECUTIVE SECRETARIAT
562686

RECEIVED
APR 10 2002
NATIONAL PARK SERVICES
CORRESPONDENCE

Erie Co.

The Town of Cheektowaga

KEVIN G. SCHENK
DEPUTY TOWN ATTORNEY

JENNIFER A. RUNFOLA
DEPUTY TOWN ATTORNEY

MICHAEL J. STACHOWSKI
TOWN ATTORNEY

CHEEKTOWAGA TOWN HALL
3301 BROADWAY
CHEEKTOWAGA, NEW YORK 14227
(716) 686-3457
FAX (716) 686-3997

CONSTANCE M. PAOLETTI
ADMINISTRATIVE ASSISTANT - LEGAL

March 25, 2002

Hon. Jack Quinn
Congressman, 30th District
403 Main Street
Brisbane Building, Room 240
Buffalo, NY 14203

Re: **War of 1812 Cemetery**
Designation as a Historic Landmark on the NYS and Federal Registries

Dear Congressman Quinn:

The Town of Cheektowaga is the owner of property known as the War of 1812 Cemetery located on Aero Drive in the Town of Cheektowaga. We have been working with the New York Sate Office of Parks, Recreation and Historic Preservation in attempting to obtain landmark status for this cemetery and are getting close to such goal. We are now provisionally listed on the National Register of Historic Places.

It would be very helpful if your office could, in writing, support our request. Please send your letter of support to me on this matter.

If you have any questions on this matter, please feel free to call me.

Very truly yours,
TOWN OF CHEEKTOWAGA
DEPARTMENT OF LAW

Kevin G. Schenk

Kevin G. Schenk
Deputy Town Attorney

Office of The Town Clerk
Richard M. Moleski
Town Clerk
686-3434
Mary F. Holtz
1st Deputy
686-3433
Vickie L. Dankowski
2nd Deputy
686-3982

RECEIVED
MAY - 9 2002

Growing In A New Direction
TOWN OF CHEEKTOWAGA
Erie County, New York

May 7, 2002

Claire Ross
Peebles Island
PO Box 189
Waterford, New York 12188-0189

Ms. Ross:

Enclosed please find a certified copy of the resolution(s) adopted by the Cheektowaga Town Board on May 6, 2002 regarding:

1. Authorize Town Historian to formally file National Registry of Historic Places Registration Form (War of 1812 Cemetery)

BY ORDER OF THE TOWN BOARD
OF THE TOWN OF CHEEKTOWAGA
ERIE COUNTY, NEW YORK

Richard M. Moleski
Town Clerk

RMM/lb
Enc(s).

Town Hall, 3301 Broadway Street ✦ Cheektowaga, New York 14227-1088

[redacted]

State Of New York
Erie County ss:
Office Of The Clerk Of The
Town of Cheektowaga

(Seal)

This is to certify that I, *RICHARD M. MOLESKI*, Clerk of the *Town of Cheektowaga, in the said County of Erie,* have compared the foregoing copy of resolution with the original resolution now on file at this office, and which was passed by the Town Board of the *Town of Cheektowaga*

in said *County of Erie,* on the 6th day of May, 2002, and that the same is a correct and true transcript of such original resolution and the whole thereof.

In Witness Whereof, I have hereunto set my hand and affixed the seal of said Town this 7th day of May, 2002.

Richard M. Moleski
CLERK OF THE TOWN BOARD, TOWN OF CHEEKTOWAGA, N.Y.

[redacted]

EXTRACTS FROM MINUTES OF CHEEKTOWAGA TOWN BOARD

At a regular meeting of the Town Board of the Town of Cheektowaga, Erie County, New York held at the Town Hall, corner of Broadway and Union Road, in said Town on the 6th day of May, 2002 at 7:00 P.M., there were:

PRESENT: Supervisor Dennis H. Gabryszak
Councilmember Patricia A. Jaworowicz
Councilmember William P. Rogowski
Councilmember Thomas M. Johnson, Jr.
Councilmember Jeff Swiatek
Councilmember James J. Jankowiak
Councilmember Thomas J. Mazur

ABSENT:

Motion by Councilmember Johnson **Seconded by** Councilmember Swiatek

WHEREAS, on May 8, 1985, the Town of Cheektowaga became the owner of property known as the War of 1812 Cemetery located on Aero Drive in the Town of Cheektowaga, and

WHEREAS, the Town of Cheektowaga believes this cemetery is historically significant for both the nation and the Western New York area, and serves as a tribute to American and British soldiers who died during the War of 1812, and

WHEREAS, the Town has sponsored annual commemorative ceremonies for many years, both in honor of the combatants who fought during the War of 1812 and in recognition of the friendship that has since persisted between the United States, Great Britain and Canada, and

WHEREAS, in the late 1990's, the Town of Cheektowaga obtained forms from the New York State Office of Parks, Recreation and Historic Preservation for the nomination of this cemetery on the State and Federal Registers of Historic Places, and

WHEREAS, since such time, Town Historian Mary F. Holtz, in concert with Claire L. Ross, Program Analyst for the Historic Field Services Bureau of the New York State Office of Parks, Recreation and Historic Preservation, has spent countless hours researching archives and other historic records throughout Western New York, in Albany and in Washington, D.C. to learn more about the War of 1812 and this cemetery, and

WHEREAS, in 2001, the Town retained a consultant to perform a Geophysical Survey to confirm the extent of the burials and the true boundaries of the cemetery; and said confirming survey was completed and filed with the New York State Office of Parks, Recreation and Historic Preservation, and

WHEREAS, on April 24, 2002, Town Historian Mary F. Holtz submitted the National Register of Historic Places Registration Form for the War of 1812 Cemetery to the NYS Office of Parks, Recreation and Historic Preservation for review, and

WHEREAS, Claire L. Ross, Historic Field Services Bureau Program Analyst, has informed the Town that she and others in her office will assist the Town of Cheektowaga with obtaining landmark status for the War of 1812 Cemetery by making refinements to the Registration Form to enable it to be submitted at the semi-annual meeting of the Historic Preservation Field Services Bureau in July or September of 2002 for nomination as a historic landmark, NOW, THEREFORE, BE IT

RESOLVED, that this Town Board hereby formally files the aforesaid Registration Form prepared by Town Historian Mary F. Holtz on behalf of the Town with the New York State Department of Parks, Recreation and Historic Preservation, and BE IT FURTHER

RESOLVED, that this Town Board hereby expresses its gratitude to Town Historian Mary F. Holtz for the many hours she has spent researching the historic significance of the War of 1812 Cemetery and in preparation of the Nomination Form, and BE IT FURTHER

RESOLVED, that this Town Board hereby also expresses its gratitude to Claire Ross and the NYS Office of Parks, Recreation and Historic Preservation for the invaluable assistance they have provided to the Town of Cheektowaga in preparing the Registration Form and research relating thereto, and BE IT FURTHER

RESOLVED, that this Town Board hereby thanks Congressman Jack Quinn, Assemblyman Paul A, Tokasz and Senator William T. Stachowski for their support of the Town's application for landmark status for the War of 1812 Cemetery, and BE IT FURTHER

RESOLVED, that this Town Board hereby directs the Town Clerk to forward certified copies of this resolution to Town Historian Mary F. Holtz, Claire Ross, the New York State Office of Parks, Recreation and Historic Preservation, Congressman Quinn, Assemblyman Tokasz and Senator Stachowski.

Upon roll call...

Supervisor Gabryszak	**Voting**	AYE
Councilmember Jaworowicz	**Voting**	AYE
Councilmember Rogowski	**Voting**	AYE
Councilmember Johnson	**Voting**	AYE
Councilmember Swiatek	**Voting**	AYE
Councilmember Jankowiak	**Voting**	AYE
Councilmember Mazur	**Voting**	AYE

AYES: 7
NAYES: 0
ABSENT: 0

F:\Attorney\Resolutions\02May6\War of 1812 cemetery reso.doc

The Town of Cheektowaga

CHEEKTOWAGA TOWN HALL
3301 BROADWAY
CHEEKTOWAGA, NEW YORK 14227
(716) 686-3465
DHG@NYSNET.NET

DENNIS H. GABRYSZAK
TOWN SUPERVISOR

July 18, 2002

JUL 2 2 2002

Claire L. Ross, Program Analyst
NYS Office of Parks, Recreation & Historic Preservation
Historic Field Services Bureau
Peebles Island
P.O. Box 189
Waterford, NY 12188-0189

Re: **Designation of War of 1812 Cemetery as a Historic Landmark On State and Federal Registers of Historic Places**

Dear Ms. Ross:

The Town of Cheektowaga is the owner of property known as the War of 1812 Cemetery on Aero Drive in the Town of Cheektowaga.

As Supervisor of the Town of Cheektowaga, I fully support the nomination and listing of the War of 1812 Cemetery on the New York State and Federal Registers of Historic Places. The Cheektowaga Town Board has gone on record requesting the historic designation of this property.

We believe that this cemetery has historic significance as a burial place for the soldiers who fought in this area during the War of 1812. This cemetery serves as a tribute to the many American soldiers who died during the War of 1812. Since the War of 1812 is important in the history of the United States, we feel it is worthy of nomination. We are very anxious to obtain landmark status for this cemetery and encourage you to do whatever is necessary to expedite the listing of this property on the State and Federal Registers of Historic Places.

Should you need any further information from me on this matter, please feel free to contact me.

Very truly yours,

TOWN OF CHEEKTOWAGA

Dennis H Gabryszak

Dennis H. Gabryszak
Supervisor

Cc: Town Board
Town Attorney

BIBLIOGRAPHY

Ashburn, Percy M. (1929). *A History of the Medical Department of the United States Army.* Houghton Mifflin, *448*, pp. 33-34.

Ashburn, Percy M. (1930). *American Army Hospitals of the Revolution and the War of 1812.*

Bulletin of the Johns Hopkins Hospital, *46*, p.32.

Bigelow, Rev. Albert Bigelow. The Early Firm of Juba Storrs & Co."

Bilow, Jack. A War of 1812 Death Register

Bingham, Robert W. Names of the United States Soldiers Buried in the Old Military Cemetery in Williamsville, New York Niagara Frontier Miscellany xxxiv, Buffalo Historical Society, 1947, p. 106.

Bingham, Robert W. *The History of Williamsville.* Niagara Frontier Miscellany xxxiv, Buffalo Historical Society, 1947, p. 97.

Cruikshank, Capt. E. ed., (1897) The Documentary History of the Campaign on the Niagara Frontier in 1814. Part II Published by Printed at the Tribune Office, Welland, Ontario, Canada

Devoy, John (1896). *A History of Buffalo and Niagara Falls*, The Buffalo Times.

Elliott , Charles Winslow (1937) Winfield Scott The Soldier and the Man

Geier, Orr and Reeves. 2006, *Huts and History*, University Press of Florida.

Ketchum, William (1864). *An Authentic and Comprehensive History of Buffalo, with Some Account of Its Early Inhabitants Both Savage and Civilized* , pp. 256, 257.

Letter dated December 12, 1812, Lt. Patrick McDonough to his parents

Litt, Williams & Whitehorne (1993) Death at Snake Hill Secrets from a War of 1812 Cemetery Dundurn Press

Livsey, Karen E. (1991). Western New York Land Transactions 1804 – 1824, Genealogical Publishing Co.

Mann, James M.D.A.A.S, (1816) Medical Sketches of the Campaigns of 1812, 13, 14, Printed by H. Mann & Company

Myers, Mordecai (2010). *Reminiscences, 1780-1814, Including Incidents in the War of 1812,* Vol. 1 Washington, 1900, *Nabu Press.*

Norton, Jacob Porter. A Yankee on the Niagara Frontier 1967

BIBLIOGRAPHY

Pfeiffer, Susan & Williamson, Ron, (1991). *Snake Hill: An Investigation of a Military Cemetery from the War of 1812*, Dundurn Press, p. 40.

Publications of the Buffalo Historical Society, Volume V

Whitehorne, Joseph. (1992). *While Washington Burned The Battle for Fort Erie*, The Nautical & Aviation Publishing Company of America, pp.50, 55.

Young, Sue Miller. (1965). *A History of the Town of Amherst, New York: 1818-1965*, Town Board of Amherst, NY.

SOURCES

Buffalo & Erie County Public Library

Cheektowaga Historical Association

Cheektowaga Town Historian Maureen E. Gleason

Claire L. Ross Program Analyst
NYS Office of Parks, Recreation and Historic Preservation

Discover GIS Data NY

Erie County, New York
County Deeds of Records

Fairchild Aerial Surveys 1927

Google Maps & Google Earth

Holland Land Office Museum
Executive Director: Ryan Duffy

Niles' Weekly Register Article from January 10, 1815

SOURCES

The Amherst Bee, Thursday, July 21, 1898

The Amherst Bee, Thursday, January 5, 1899

The Amherst Bee, Thursday, April 6, 1939

The Amherst Bee Thursday, May 25, 1939

The Buffalo Commercial, Tuesday, November 1, 1898

The Buffalo Commercial, Wednesday, November 9, 1898

Buffalo Courier, Sunday, September 7, 1924

Buffalo Courier Express, Tuesday, November 8, 1898

The Buffalo Enquirer, Tuesday, June 26, 1923

The Buffalo News, Monday, November 7, 1898

The Buffalo News, Tuesday, April 25, 1939

The Buffalo Review, Wednesday, November 2, 1898

The Buffalo Times, Sunday, February 6, 1938

The Buffalo History Museum
Director of Library & Archives: Cynthia Van Ness

US ARMY MEDICAL DEPARTMENT Office of Medical History

War of 1812 Cemetery Timeline
Courtesy of David Przepior

ABOUT THE AUTHOR

J. A. Buscaglia's roots run deep in Western New York, he found himself drawn to the whispers of history echoing through its corners. It was this childhood fascination that ignited his lifelong passion for uncovering the tales of Western New York's pioneers and the tumultuous era of the War of 1812. His expertise primarily lies in historical research, with a focus on the early settlers of Western New York, the Holland Land Company, and the events surrounding the War of 1812 on the Niagara Frontier.

In his debut work, "Searching for Jonas Williams," Buscaglia delves into a mystery that has long plagued historians-the final resting place of Williamsville's founder. When human remains were unearthed in 1969 during construction in the Village of Williamsville, speculation ran rampant that they belonged to Jonas Williams himself. Yet, Buscaglia's meticulous research challenges this assumption, offering compelling evidence that Williams' true grave remains undiscovered. With divergent opinions clouding the truth, Buscaglia courageously shares his own insights, acknowledging the sensitive nature of the subject and urging respect for private property.

His subsequent book, "Williamsville and the War of 1812," unravels another chapter of local history. Amidst the chaos of war, Williamsville played a pivotal role, hosting thousands of soldiers and witnessing both triumph and tragedy. From makeshift barracks to the somber grounds of a military hospital, the landscape bore witness to untold suffering and sacrifice. Buscaglia's narrative weaves together accounts of heroism and hardship, shedding light on forgotten landmarks and honoring the memories of those who served.

In his third book, "Ancient Forts, Mounds, and Burial Grounds of WNY," Buscaglia ventures further back in time. Inspired by the writings of T. Apoleon Cheney and others, he embarks on a journey to unearth the remnants of a forgotten civilization. Drawn to the enigmatic earthworks and burial sites scattered throughout Western New York, Buscaglia seeks to preserve their legacy and unravel the mysteries they hold. With reverence for the sacredness of these sites, he emphasizes the importance of obtaining proper permission before exploring them, mindful of their significance to both the past and present.

ALSO BY J A BUSCAGLIA

FINDING JONAS WILLIAMS: THE SEARCH FOR JONAS WILLIAMS' GRAVE

In 1969 human bones were unearthed by workmen at a new office building in the Village of Williamsville New York and historians believed it was the unmarked grave of Jonas Williams founder of Williamsville, New York.

At the time members of the Amherst Historical Society and Town Historian, were willing to be certain that Williamsville had found its forbearer.

My extensive research will show that location is not the grave of Williamsville founder Jonas Williams and that his remains still laid undiscovered until now.

With there being such a diversity of opinion on the burial location of Jonas Williams and whether his remains have been discovered, I have decided to write my thoughts and observations on the topic. So, I am revealing all the information regarding this subject that I have been able to obtain from any source in order to determine the site, which I believe to be the location of Jonas Williams remains.

This is just my opinion from my historical research and my observations. It is not an academic paper but a first-person account of my findings.

Please be mindful that these are human remains and that the location I will reveal is on private property and trespassing is a violation of the law. I in no way endorse trespassing or breaking the law in any form.

WILLIAMSVILLE AND THE WAR OF 1812: "GA-DA-YA-DEH" A PLACE OF MISERY

During the War of 1812 an army of 5,000 to 6,000 men were quartered at Williamsville, New York which was also known as Eleven Mile Creek, the Evans House became the temporary headquarters of Generals Smyth, Brown and Scott, and a military hospital was established.

Long before the settlers came, the Senecas called the site of present-day Williamsville, New York "Ga-da-ya-deh" which means a place of misery, maybe they foresaw what was to come, or maybe something had happened in their past? One thing is certain: between the years of 1812 and 1815, there was plenty of misery from the executions of deserters on the parade ground on the site of present-day Saints Peter and Paul's church to the unimaginable suffering in the general hospital which accommodated from 300 to 400 sick and a somewhat larger number of wounded in the present day Lehn Springs development.

The only locations which are known of the Army's encampment are the locations of the War of 1812 Cemetery and the Evans house.

The diversity of opinion on the locations of the War of 1812 Barracks and Hospital, being so great that I have written my thoughts and observations on the topic. Hence, I am revealing all the information regarding this subject that I have been able to gather from any source in order to determine the sites. This is simply my opinion from my historical research and observations. It is not an academic paper, but a first-person account of my findings.

ANCIENT FORTS, MOUNDS, AND BURIAL GROUNDS OF WNY

Buscaglia ventures further back in time. Inspired by the writings of T. Apoleon Cheney and others, he embarks on a journey to unearth the remnants of a forgotten civilization. Drawn to the enigmatic earthworks and burial sites scattered throughout Western New York, Buscaglia seeks to preserve their legacy and unravel the mysteries they hold. With reverence for the sacredness of these sites, he emphasizes the importance of obtaining proper permission before exploring them, mindful of their significance to both the past and present.

Made in the USA
Middletown, DE
05 November 2026

36545326R10077